MTEL Chemistry

12 Teacher Certification Exam

By: Sharon Wynne, M.S
Southern Connecticut State University

"And, while there's no reason yet to panic, I think it's only prudent that we make preparations to panic."

XAMonline, INC.

Boston

To obtain permission(s) to use the material from this work for any purpose including workshops or seminars, please submit a written request to:

XAMonline, Inc.
21 Orient Ave.
Melrose, MA 02176
Toll Free 1-800-509-4128
Email: info@xamonline.com
Web www.xamonline.com
Fax: 1-781-662-9268

Library of Congress Cataloging-in-Publication Data

Wynne, Sharon A.
    Chemistry 12: Teacher Certification / Sharon A. Wynne. -2nd ed.
    ISBN 978-1-58197-883-4
    1. Chemistry 12.   2. Study Guides.   3. MTEL
    4. Teachers' Certification & Licensure.   5. Careers

Disclaimer:

The opinions expressed in this publication are the sole works of XAMonline and were created independently from the National Education Association, Educational Testing Service, or any State Department of Education, National Evaluation Systems or other testing affiliates.

Between the time of publication and printing, state specific standards as well as testing formats and website information may change that is not included in part or in whole within this product. Sample test questions are developed by XAMonline and reflect similar content as on real tests; however, they are not former tests. XAMonline assembles content that aligns with state standards but makes no claims nor guarantees teacher candidates a passing score. Numerical scores are determined by testing companies such as NES or ETS and then are compared with individual state standards. A passing score varies from state to state.

Printed in the United States of America   œ-1

MTEL: Chemistry 12
ISBN: 978-1-58197-883-4

Massachusetts Tests for Educator Licensure®

MARC SHELIKOFF has met the qualifying score on the following test(s) as of March 4, 2006:
  12 Chemistry

Test Date: March 4, 2006

See reverse side for an explanation of how to read your score report.

MARC SHELIKOFF

Your scores have been reported to the Massachusetts Department of Education.

---

12 Chemistry

Your Score: 94	Minimum Qualifying Score: 70	Status: Met the Qualifying Score

Number of Questions	Subarea Name	Graphic Display
1 to 10	The Nature of Chemical Inquiry...	
11 to 20	Matter and Atomic Structure...	
11 to 20	Energy/Chemical Bonds/Molecular Struct............................	
11 to 20	Chemical Reactions..	
1 to 10	Quantitative Relationships..	
11 to 20	Chemistry, Society, and the Environment.............................	
2	Open-Response Items..	

About The MTEL Chemistry Test (12)

The MTEL Chemistry test is designed to assess the candidate's knowledge of the subject matter required for the Massachusetts Chemistry Teacher certificate. This subject matter knowledge is delineated in the Massachusetts Department of Education *Regulations for the Certification of Educational Personnel in Massachusetts* (April 1995), 603 C.M.R. 7.12, "Competencies for Specific Certificates," Section (10) (a) 2. "Competency I: Subject Matter Knowledge."

The test assesses the candidate's proficiency and depth of understanding of the subject at the level required for a baccalaureate major, according to Massachusetts standards. Candidates are typically nearing completion of or have completed their undergraduate work when they take the test.

The test comprises 80 multiple-choice items and two (2) open-response items. The multiple-choice section will account for 75 percent of the points available and the constructed-response section will account for 25 percent of the points available.

The multiple-choice items on the test cover the subareas as indicated in the chart below. The open-response items, for which candidates are typically asked to prepare a written response or to solve a problem, may relate to topics covered in any of the subareas and will typically require breadth of understanding of the chemistry field and the ability to relate concepts from different aspects of the field. Each open-response item is expected to take a typical examinee response time of about 45-60 minutes.

Subareas	Approximate Number of Multiple-Choice Items	Number of Open-Response Items
I. The Nature of Chemical Inquiry	10–12	2
II. Matter and Atomic Structure	15–17	
III. Energy, Chemical Bonds, and Molecular Structure	13–15	
IV. Chemical Reactions	13–15	
V. Quantitative Relationships	10–12	
VI. Interactions of Chemistry, Society, and the Environment	13–15	

Table of Contents

Study and Testing Tips vi

Periodic Table of the Elements x

SUBAREA I.   THE NATURE OF CHEMICAL INQUIRY

Competency 0001   Understand the nature of scientific inquiry, scientific processes, and the role of observation and experimentation in science. 1

Competency 0002   Understand the processes of gathering, organizing, reporting, and interpreting scientific data in the context of chemistry investigations. 7

Competency 0003   Understand principles and procedures of measurement used in chemistry. 20

Competency 0004   Understand proper, safe, and legal use of equipment, materials, and chemicals used in chemistry investigations 28

SUBAREA II.   MATTER AND ATOMIC STRUCTURE

Competency 0005   Understand the concept of matter, and analyze chemical and physical properties of and changes in matter 49

Competency 0006   Understand the various models of atomic structure, the principles of quantum theory, and the properties and interactions of subatomic particles. 55

Competency 0007   Understand the organization of the periodic table. 63

Competency 0008   Understand the kinetic theory, the nature of phase changes, and the gas laws. 75

Competency 0009   Apply the conventions of chemical notation and representations. 82

Competency 0010   Understand the process of nuclear transformation. 94

SUBAREA III. ENERGY, CHEMICAL BONDS, AND MOLECULAR STRUCTURE

Competency 0011   Understand the principles of thermodynamics and calorimetry 108

Competency 0012   Understand energy relationships in chemical bonding and chemical reactions. 117

Competency 0013   Understand the types of bonds between atoms (including ionic, covalent, and metallic bonds), the formation of these bonds, and properties of substances containing the different bonds 119

Competency 0014   Understand types and characteristics of molecular interaction and properties of substances containing different types of interactive forces between molecules 126

Competency 0015   Understand the nomenclature and structure of organic compounds 133

SUBAREA IV.   CHEMICAL REACTIONS

Competency 0016   Understand factors that affect reaction rates and methods of measuring reaction rates... 147

Competency 0017   Understand the principles of chemical equilibrium. 158

Competency 0018   Understand the theories, principles, and applications of acid-base chemistry 169

Competency 0019   Understand redox reactions and electrochemistry. 181

Competency 0020   Understand the nature of organic reactions. 193

SUBAREA V.   QUANTITATIVE RELATIONSHIPS

Competency 0021   Understand the mole concept. 200

Competency 0022   Understand the relationship between the mole concept and chemical formulas. 203

Competency 0023   Understand the quantitative relationships expressed in chemical equations 207

Competency 0024   Understand the properties of solutions and colloidal suspensions, and analyze factors that affect solubility 216

SUBAREA VI.   INTERACTIONS OF CHEMISTRY, SOCIETY, AND THE ENVIRONMENT

Competency 0025   Understand the historical and contemporary contexts of the study of chemistry. 226

Competency 0026   Understand the chemistry of practical processes and applications of chemical theory to other scientific disciplines. 233

Competency 0027   Understand the applications of nuclear reactions 234

Competency 0028   Understand factors and processes related to the release of chemicals into the environment.......................................238

Competency 0029   Understand the interrelationships among chemistry, society, technology, and other disciplines. 240

Sample Test..242

Answer Key 267

Rationales with Sample Questions 268

Sample Open-Response Questions 334

Sample Open-Response Answers..336

Great Study and Testing Tips!

What to study in order to prepare for the subject assessments is the focus of this study guide but equally important is *how* you study.

You can increase your chances of truly mastering the information by taking some simple, but effective steps.

Study Tips:

1. <u>Some foods aid the learning process</u>. Foods such as milk, nuts, seeds, rice, and oats help your study efforts by releasing natural memory enhancers called CCKs (*cholecystokinin*) composed of *tryptophan*, *choline*, and *phenylalanine*. All of these chemicals enhance the neurotransmitters associated with memory. Before studying, try a light, protein-rich meal of eggs, turkey, and fish. All of these foods release the memory enhancing chemicals. The better the connections, the more you comprehend.

Likewise, before you take a test, stick to a light snack of energy boosting and relaxing foods. A glass of milk, a piece of fruit, or some peanuts all release various memory-boosting chemicals and help you to relax and focus on the subject at hand.

2. <u>Learn to take great notes</u>. A by-product of our modern culture is that we have grown accustomed to getting our information in short doses (i.e. TV news sound bites or USA Today style newspaper articles.)

Consequently, we've subconsciously trained ourselves to assimilate information better in <u>neat little packages</u>. If your notes are scrawled all over the paper, it fragments the flow of the information. Strive for clarity. Newspapers use a standard format to achieve clarity. Your notes can be much clearer through use of proper formatting. A very effective format is called the *"Cornell Method."*

Take a sheet of loose-leaf lined notebook paper and draw a line all the way down the paper about 1-2" from the left-hand edge.

Draw another line across the width of the paper about 1-2" up from the bottom. Repeat this process on the reverse side of the page.

Look at the highly effective result. You have ample room for notes, a left hand margin for special emphasis items or inserting supplementary data from the textbook, a large area at the bottom for a brief summary, and a little rectangular space for just about anything you want.

3. Get the concept then the details. Too often we focus on the details and don't gather an understanding of the concept. However, if you simply memorize only dates, places, or names, you may well miss the whole point of the subject.

A key way to understand things is to put them in your own words. If you are working from a textbook, automatically summarize each paragraph in your mind. If you are outlining text, don't simply copy the author's words.

Rephrase them in your own words. You remember your own thoughts and words much better than someone else's, and subconsciously tend to associate the important details to the core concepts.

4. Ask Why? Pull apart written material paragraph by paragraph and don't forget the captions under the illustrations.

Example: If the heading is "Stream Erosion", flip it around to read "Why do streams erode?" Then answer the questions.

If you train your mind to think in a series of questions and answers, not only will you learn more, but it also helps to lessen the test anxiety because you are used to answering questions.

5. Read for reinforcement and future needs. Even if you only have 10 minutes, put your notes or a book in your hand. Your mind is similar to a computer; you have to input data in order to have it processed. *By reading, you are creating the neural connections for future retrieval.* The more times you read something, the more you reinforce the learning of ideas.

Even if you don't fully understand something on the first pass, *your mind stores much of the material for later recall.*

6. Relax to learn so go into exile. Our bodies respond to an inner clock called biorhythms. Burning the midnight oil works well for some people, but not everyone.

If possible, set aside a particular place to study that is free of distractions. Shut off the television, cell phone, pager and exile your friends and family during your study period.

If you really are bothered by silence, try background music. Light classical music at a low volume has been shown to aid in concentration over other types. Music that evokes pleasant emotions without lyrics are highly suggested. Try just about anything by Mozart. It relaxes you.

7. <u>**Use arrows not highlighters.**</u> At best, it's difficult to read a page full of yellow, pink, blue, and green streaks. Try staring at a neon sign for a while and you'll soon see that the horde of colors obscure the message.

A quick note, a brief dash of color, an underline, and an arrow pointing to a particular passage is much clearer than a horde of highlighted words.

8. <u>**Budget your study time.**</u> Although you shouldn't ignore any of the material, *allocate your available study time in the same ratio that topics may appear on the test.*

Testing Tips:

1. **Get smart, play dumb.** **Don't read anything into the question.** Don't make an assumption that the test writer is looking for something else than what is asked. Stick to the question as written and don't read extra things into it.

2. **Read the question and all the choices *twice* before answering the question.** You may miss something by not carefully reading, and then re-reading both the question and the answers.

If you really don't have a clue as to the right answer, leave it blank on the first time through. Go on to the other questions, as they may provide a clue as to how to answer the skipped questions.

If later on, you still can't answer the skipped ones . . . *Guess.* The only penalty for guessing is that you *might* get it wrong. Only one thing is certain; if you don't put anything down, you will get it wrong!

3. **Turn the question into a statement.** Look at the way the questions are worded. The syntax of the question usually provides a clue. Does it seem more familiar as a statement rather than as a question? Does it sound strange?

By turning a question into a statement, you may be able to spot if an answer sounds right, and it may also trigger memories of material you have read.

4. **Look for hidden clues.** It's actually very difficult to compose multiple-foil (choice) questions without giving away part of the answer in the options presented.

In most multiple-choice questions you can often readily eliminate one or two of the potential answers. This leaves you with only two real possibilities and automatically your odds go to Fifty-Fifty for very little work.

5. **Trust your instincts.** For every fact that you have read, you subconsciously retain something of that knowledge. On questions that you aren't really certain about, go with your basic instincts. **Your first impression on how to answer a question is usually correct.**

6. **Mark your answers directly on the test booklet.** Don't bother trying to fill in the optical scan sheet on the first pass through the test.

Just be very careful not to mis-mark your answers when you eventually transcribe them to the scan sheet.

7. **Watch the clock!** You have a set amount of time to answer the questions. Don't get bogged down trying to answer a single question at the expense of 10 questions you can more readily answer.

Periodic Table of the Elements

Group	1	2	3	4	5	6	7	8	9	10	11	12	13	14	15	16	17	18
	I	II											III	IV	V	VI	VII	VIII
Period	IA	IIA	IIIB	IVB	VB	VIB	VIIB	VIIIB	VIIIB	VIIIB	IB	IIB	IIIA	IVA	VA	VIA	VIIA	VIIIA
1	hydrogen 1 H 1.0079																	helium 2 He 4.0026
2	lithium 3 Li 6.941	beryllium 4 Be 9.0122											boron 5 B 10.811	carbon 6 C 12.011	nitrogen 7 N 14.007	oxygen 8 O 15.999	fluorine 9 F 18.998	neon 10 Ne 20.180
3	sodium 11 Na 22.990	magnesium 12 Mg 24.305											aluminum 13 Al 26.982	silicon 14 Si 28.086	phosphorus 15 P 30.974	sulfur 16 S 32.065	chlorine 17 Cl 35.453	argon 18 Ar 39.948
4	potassium 19 K 39.098	calcium 20 Ca 40.078	scandium 21 Sc 44.956	titanium 22 Ti 47.867	vanadium 23 V 50.942	chromium 24 Cr 51.996	manganese 25 Mn 54.938	iron 26 Fe 55.845	cobalt 27 Co 58.933	nickel 28 Ni 58.693	copper 29 Cu 63.546	zinc 30 Zn 65.409	gallium 31 Ga 69.723	germanium 32 Ge 72.64	arsenic 33 As 74.922	selenium 34 Se 78.96	bromine 35 Br 79.904	krypton 36 Kr 83.798
5	rubidium 37 Rb 85.468	strontium 38 Sr 87.62	yttrium 39 Y 88.906	zirconium 40 Zr 91.224	niobium 41 Nb 82.906	molybdenum 42 Mo 95.94	technetium 43 Tc [98]	ruthenium 44 Ru 101.07	rhodium 45 Rh 102.91	palladium 46 Pd 106.42	silver 47 Ag 107.87	cadmium 48 Cd 112.41	indium 49 In 114.82	tin 50 Sn 118.71	antimony 51 Sb 121.76	tellurium 52 Te 127.60	iodine 53 I 126.90	xenon 54 Xe 131.29
6	cesium 55 Cs 132.91	barium 56 Ba 137.33	57-71 *	hafnium 72 Hf 178.49	tantalum 73 Ta 180.95	tungsten 74 W 183.84	rhenium 75 Re 186.21	osmium 76 Os 190.23	iridium 77 Ir 192.22	platinum 78 Pt 195.08	gold 79 Au 196.97	mercury 80 Hg 200.59	thallium 81 Tl 204.38	lead 82 Pb 207.2	bismuth 83 Bi 208.98	polonium 84 Po [209]	astatine 85 At [210]	radon 86 Rn [222]
7	francium 87 Fr [223]	radium 88 Ra [226]	89-103 **	rutherfordium 104 Rf [261]	dubnium 105 Db [262]	seaborgium 106 Sg [266]	bohrium 107 Bh [264]	hassium 108 Hs [277]	meitnerium 109 Mt [268]	darmstadtium 110 Ds [271]	roentgenium 111 Rg [272]							

*Lanthanoids	lanthanum 57 La 138.91	cerium 58 Ce 140.12	praseodymium 59 Pr 140.91	neodymium 60 Nd 144.24	promethium 61 Pm [145]	samarium 62 Sm 150.36	europium 63 Eu 151.96	gadolinium 64 Gd 157.25	terbium 65 Tb 158.93	dysprosium 66 Dy 162.50	holmium 67 Ho 164.93	erbium 68 Er 167.26	thulium 69 Tm 168.93	ytterbium 70 Yb 173.0	lutetium 71 Lu 174.97
**Actinoids	actinium 89 Ac [227]	thorium 90 Th 232.04	protactinium 91 Pa 231.04	uranium 92 U 238.03	neptunium 93 Np [237]	plutonium 94 Pu [244]	americium 95 Am [243]	curium 96 Cm [247]	berkelium 97 Bk [247]	californium 98 Cf [251]	einsteinium 99 Es [252]	fermium 100 Fm	mendelevium 101 Md [258]	nobelium 102 No	lawrencium 103 Lr [262]

Atomic mass values from IUPAC review (2001): http://www.iupac.org/reports/periodic_table/

"There are and can be only two ways (inductive and deductive methods) of searching into and discovering truth. The one (deductive) flies from the senses and the particulars to the most general axioms and from these (principles); the truth it takes as fore settled and immovable proceeds to judgment and to the discovery of middle axioms and this way is now in fashion. The other (inductive) derives axioms from the senses and particulars rising by a gradual and unbroken assent, that it arrives at the most general axiom last of all. This is the true way."

Sir Francis Bacon

COMPETENCY 0001   UNDERSTAND THE NATURE OF SCIENTIFIC INQUIRY, SCIENTIFIC PROGRESSES, AND THE ROLE OF OBSERVATION AND EXPERIMENTATION IN SCIENCE.

Modern science began around the late 16th century with a new way of thinking about the world. Few scientists will disagree with Carl Sagan's assertion that "science is a way of thinking much more than it is a body of knowledge" (Broca's Brain, 1979). Thus science is a process of inquiry and investigation. It is a way of thinking and acting, not just a body of knowledge to be acquired by memorizing facts and principles. This way of thinking, the scientific method, is based on the idea that scientists begin their investigations with observations. From these observations they develop a hypothesis, which is extended in the form of a prediction, and challenge the hypothesis through experimentation and thus further observations. Science has progressed in its understanding of nature through careful observation, a lively imagination, and increasing sophisticated instrumentation. Science is distinguished from other fields of study in that it provides guidelines or methods for conducting research, and the research findings must be reproducible by other scientists for those findings to be valid. It is important to recognize that scientific practice is not always this systematic. Discoveries have been made that are serendipitous and others have not started with the observation of data. Einstein's theory of relativity started with an intellectual and enquiring mind.

The Scientific Method is just a logical set of steps that a scientist goes through to solve a problem. There are as many different scientific methods as there are scientists experimenting. However, there seems to be some pattern to their work.

While an inquiry may start at any point in this method and may not involve all of the steps here is the pattern.

Observations

Scientific questions result from observations of events in nature or events observed in the laboratory. An **observation** is not just a look at what happens. It also includes measurements and records of the event. Records could include photos, drawings, or written descriptions. The observations and data collection lead to a question. In chemistry, observations almost always deal with the behavior of matter. Having arrived at a question, a scientist usually researches the scientific literature to see what is known about the question. Maybe the question has already been answered. The scientist then may want to test the answer found in the literature. Or, maybe the research will lead to a new question.

Sometimes the same observations are made over and over again and are always the same. For example, you can observe that daylight lasts longer in summer than in winter. This observation never varies. Such observations are called **laws** of nature. Probably the most important law in chemistry was discovered in the late 1700s. Chemists observed that mass neither lost nor gained in chemical reactions. This law became known as the law of conservation of mass. Explaining this law was a major topic of chemistry in the early 19th century.

Hypothesis

If the question has not been answered, the scientist may prepare for an experiment by making a hypothesis. A **hypothesis** is a statement of a possible answer to the question. It is a tentative explanation for a set of facts and can be tested by experiments. Although hypotheses are usually based on observations, they may also be based on a sudden idea or intuition.

Experiment

An **experiment** tests the hypothesis to determine whether it may be a correct answer to the question or a solution to the problem. Some experiments may test the effect of one thing on another under controlled conditions. Such experiments have two variables. The experimenter controls one variable, called the *independent variable*. The other variable, the *dependent variable*, is the change caused by changing the independent variable. In other words, a dependent variable is the factor that is measured in an experiment and independent variables are things that are changed or manipulated in an experiment.

For example, suppose a researcher wanted to test the effect of vitamin A on the ability of rats to see in dim light. The independent variable would be the dose of Vitamin A added to the rats' diet. The dependent variable would be the ability of rats to see in dim light and this could be measured by placing food at different distances.

All other factors, such as time, temperature, age, water , and other nutrients given to the rats, and similar factors, are held constant. Chemists sometimes do short experiments "just to see what happens" or to see what products a certain reaction produces. Often, these are not formal experiments. Rather they are ways of making additional observations about the behavior of matter.

 The design of chemical experiments must include every step to obtain the desired data. In other words, the design must be **complete** and it must include all required **controls**

Complete design

Familiarity with individual experiments and equipment will help you evaluate if anything is missing from the design. For data requiring a difference between two values, the experiment **must determine both values**. For data utilizing the ideal gas law, the experiment **must determine three values of P, V, n, or T** in order to determine the fourth or one value and a ratio of the other two in order to determine the fourth.

Example: In a mercury manometer, the level of mercury in contact with a reaction vessel is 70.0 mm lower than the level exposed to the atmosphere. Use the following conversion factors:

760 mm Hg=1 atm=101.325 kPa.

What additional information is required to determine the pressure in the vessel in Pa?

Solution: The barometric pressure is needed to determine vessel pressure from an open-ended manometer. A manometer reading is always a **difference** between two pressures. See **0005**. One standard atmosphere is 760 mm mercury, but on a given day at a given location, the actual ambient pressure may vary. If the barometric pressure on the day of the experiment is 104 kPa, the pressure of the vessel is:

$$104 \text{ kPa} + 70.0 \text{ mm Hg} \times \frac{101.325 \text{ kPa}}{760 \text{ mm Hg}} = 113 \text{ kPa}.$$

Controls

Experimental **controls** prevent factors other than those under study from impacting the outcome of the experiment. An **experimental sample** in a controlled experiment is the unknown to be compared against one or more **control samples**. These should be nearly identical to the experimental sample except for the one aspect whose effect is being tested.

A **negative control** is a control sample that is known to lack the effect. A **positive control** is known to contain the effect.

Positive controls of varying strengths are often used to generate a **calibration curve** (also called a **standard curve**).

When determining the concentration of a component in a mixture, an **internal standard** is a known concentration of a different substance that is added to the experimental sample. An **external standard** is a known concentration of the substance of interest. External standards are more commonly used. They are not added to the experimental sample; they are analyzed separately

Replicate samples decrease the impact of random error. A mean is taken of the results from replicate samples to obtain a best value. If one replicate is obviously inconsistent with the results from other samples, it may be discarded as an **outlier** and not counted as an observation when determining the mean. Discarding an outlier is equivalent to assuming the presence of a systematic error for that particular observation. In research, this must be done with great caution because some real-world behavior generates sporadically unusual results.

Example: A pure chemical in aqueous solution is known to absorb light at 615 nm. What controls would best be used with a spectrophotometer to determine the concentration of this chemical when it is present in a mixture with other solutes in an aqueous solution?

Solution: The other solutes may also absorb light at 615 nm. The best negative control would be an identical mixture with the chemical of interest entirely absent. Known concentrations of the chemical could then be added to the negative control to create positive controls (external standards) and develop a calibration curve of the spectrophotometer absorbance reading at 615 nm as a function of concentration. Replicate samples of each standard and of the unknown should be read.

Example: Ethanol is separated from a mixture of organic compounds by gas chromatography. The concentration of each component is proportional to its peak area. However, the chromatograph detector has a variable sensitivity from one run to the next. Is an internal standard required to determine the concentration of ethanol?

Solution: Yes. The variable detector sensitivity may only be accounted for by adding a known concentration of a chemical not found in the mixture as an internal standard to the experimental sample and control samples. The variable sensitivity of the detector will be accounted for by determining the ratio of the peak area for ethanol to the peak area of the added internal standard.

In most experiments, scientists collect quantitative data, which is data that can be measured with instruments. They also collect qualitative data, descriptive information from observations other than measurements. Interpreting data and analyzing observations are important. If data is not organized in a logical manner, wrong conclusions can be drawn. Also, other scientists may not be able to follow your work or repeat your results.

Analysis of data:
It is very important to analyze the data, to see if there are variations and patterns.

Conclusion
Finally, a scientist must draw conclusions from the experiment. A conclusion must address the hypothesis on which the experiment was based. The conclusion states whether or not the data supports the hypothesis. If it does not, the conclusion should state what the experiment *did* show. If the hypothesis is not supported, the scientist uses the observations from the experiment to make a new or revised hypothesis., Then, new experiments are planned.

Theory
When a hypothesis survives many experimental tests to determine its validity, the hypothesis may evolve into a **theory**. A theory explains a body of facts and laws that are based on the facts. A theory also reliably predicts the outcome of related events in nature. For example, the law of conservation of matter and many other experimental observations led to a theory proposed early in the 19th century. This theory explained the conservation law by proposing that all matter is made up of atoms which are never created or destroyed in chemical reactions, only rearranged. This atomic theory also successfully predicted the behavior of matter in chemical reactions that had not been studied at the time. As a result, the atomic theory has stood for 200 years with only small modifications.

A theory also serves as a scientific **model**. A model can be a physical model made of wood or plastic, a computer program that simulates events in nature, a mathematical model or simply a mental picture of an idea. A model illustrates a theory and explains nature. In your chemistry course, you will develop a mental (and maybe a physical) model of the atom and its behavior. Outside of science, the word theory is often used to describe someone's unproven notion about something. In science, theory means much more. It is a thoroughly tested explanation of things and events observed in nature.

The test of the hypothesis may be observations of phenomena or a model may be built to examine its behavior under certain circumstances.

A theory is open and is subject to experimentation. Experiments may establish a theory or may reject it.

Theories provide a framework to explain the **known** information of the time, but are subject to constant evaluation and updating. There is always the possibility that new evidence will conflict with a current theory.

Some examples of theories that have been rejected because they are now better explained by current knowledge:

Theory of Spontaneous Generation
Inheritance of Acquired Characteristics
The Blending Hypothesis

Some examples of theories that were initially rejected because they fell outside of the accepted knowledge of the time, but are well-accepted today due to increased knowledge and data include:

The sun-centered solar system
Warm-bloodedness in dinosaurs
The germ-theory of disease
Continental drift

A Law is a naturally occurring phenomenon like the law of gravity .

Law is defined as: a statement of an order or relation of phenomena that so far as is known is invariable under the given conditions. Everything we observe in the universe operates according to known natural laws.

- If the truth of a statement is verified repeatedly in a reproducible way then it can reach the level of a natural law.
- Some well know and accepted natural laws of science are:

1. The First Law of Thermodynamics

2. The Second Law of Thermodynamics

3. The Law of Cause and Effect

4. The Law of Biogenesis

5. The Law of Gravity

The **atomic/molecular weights of atoms or molecules** in a mixture are determined with a **mass spectrometer**. The sample is vaporized, this gas is ionized, and these ions are deflected towards a magnet that separates them according to their mass. There are many specialized applications of mass spectrometry so there are dozens of variations to the process. Mass spectrometry is used to determine the ratio of $^2H/^1H$ in water, for ^{14}C dating, and to characterize polymers and biological molecules with molecular weights of over a million.

Perhaps what distinguishes the sciences from other fields is our insistence that models be based upon experimental evidence. Nowhere is this dependence upon evidence more appreciated than in the scientific community.

You should not be too surprised to discover an unusual emphasis on the collection of evidence and the expectation that there will be close ties between such evidence and your conclusions. Therefore, data collected in an experiment or investigation needs to be displayed in a way that is easy to understand, see connections and determine validity. This is most effectively done as a data table. From the data table below, it is easy to see that the temperature of the water is increasing as time is progressing.

When displaying measurements in a data table follow these simple rules:

- Make a table of vertical columns for the variables. Record the independent variable (IV) in the first column, the dependent variable (DV) in the second; reserve the third for any derived quantity.
- Label the data table with a title that gives information about how the measurements were collected.

The Rate of Temperature Increase of Water Upon Heating	
Time Interval (s)	Temperature of water (ºC)
0	22.5
10	25.8
20	27.7
30	30.3
40	32.5
50	35.5
60	42.6
70	48.2
80	51.1

- Label all columns and rows in the data table with information as to what the measurement is and the units used.

- Record the values of the independent variable from smallest to largest.

- Center numbers and align decimal points.

- Calculate the derived quantities and enter the values into the table.

Historically raw data, measured values, appeared in data tables and calculated values were kept separate. With the advent of computer technology, computer-based data acquisition and manipulation, this is no longer the case. Calculated values are shown next to raw data values and delineated by column and row labels.

How ever, more often than not, the data is compiled into graphs. Graphs help scientists visualize and interpret the variation in data. Depending on the nature of the data, there are many types of graphs. Bar graphs, pie charts and line graphs are just a few methods used to pictorially represent numerical data.

A graph pictorially shows the relationship between two variables, the independent variable and the dependent variable. The **independent variable** is the variable you control. The **dependent variable** is the one that changes in response to the independent variable. A graph is set up so the values for the independent variable are found along the x- axis (horizontal) and the dependent variable values are found along the y- axis (vertical).

While there are several variations of each, the three basic types of graphs are line, bar, and pie. The type of graph used depends upon the type of observations and measurements.

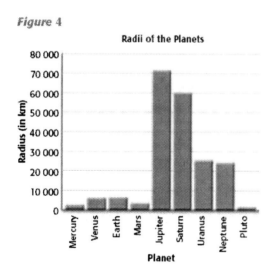

Figure 4

Radii of the Planets

Bar graphs are the most appropriate graphs for depicting discrete data. *Discrete data* are categorical or counted (for example- gender, months of the year, countries, types of seeds germinated).

If the intervals between the data do not have meaning, like brand name of fertilizers, a bar graph is the best choice.

Bar graphs in which the rectangles are arranged horizontally. The length of each rectangle represents its value. Bar graphs are sometimes referred to as histograms.

Bar graphs best show:

• data series with no natural order.

Bar graphs are good for looking at differences amongst similar things. If the data are a time series, a carefully chosen column graph is generally more appropriate but bar graphs can be used to vary a presentation when many column graphs of time series are used. One advantage of bar graphs is that there is greater horizontal space for variable descriptors because the vertical axis is the category axis.

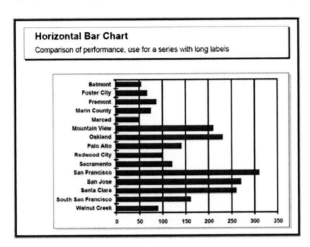

Line graphs are generally used to plot *continuous data* (measurements associated with a standard scale or continuum). Line graphs allow us to interpolate the values of points not directly measured and allow for inference /prediction of future events.

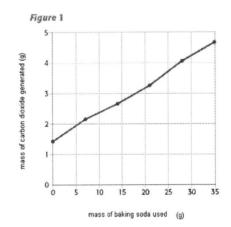

Line graphs show data points connected by lines; different series are given different line markings (for example, dashed or dotted) or different tick marks. Line graphs are useful when the data points are more important than the transitions between them. They best show:

• the comparison of long series

• a general trend is the message.

Line graphs are good for showing trends or changes over time.

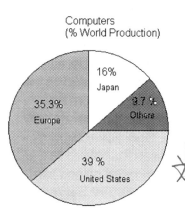

Computers
(% World Production)

Pie graphs are also known as pie charts. They are typically used to indicate how different parts make up a whole. A pie chart is a circle with radii connecting the center to the edge. The area between two radii is called a slice. Data values are proportionate to the angle between the radii.

Pie charts best show:

• parts of a whole

Be careful of too many slices since they result in a cluttered graph. Six slices are as many as can be handled on one pie.

Graph construction is more difficult than table construction, however using your past experience, the rules and tips below, you should soon be a graphing expert.

- Always use graph paper or a computer program to draw a graph.
- Graphs contain two axes. The horizontal or x-axis is used for recording the independent variable while the y-axis or vertical axis is for the dependent variable.
- Use a ruler to draw the x and y- axis.
- Use a consistent scale that is easy to work with. i.e., increasing by 1 unit, 2 units or 5 or 10 units or a multiple of 10 like 0.1, 0.01, 100 or 1000.
- Select a scale that utilizes the greatest portion of the graph paper. Mark that scale on the axis with "tick" marks.
- Label each axis with the measurement and the units they represent.
- Plot your data points
- Draw a smooth curve (or line) to represent the data.
- Title the graph so that the title represents the data and how it was collected.
- If more than one curve is plotted on the same axis, include a legend that identifies each curve. Use different colors or symbols for each curve and include a legend.
- If you determine the slope of the curve, circle the two points on the line that you are using in the calculation.
- Write the coordinates of the points next to each circle.

The interpretation of data and construction and interpretation of graphs are central practices in science. Graphs are effective visual tools which relay information quickly and reveal trends easily. While there are several different types of graphical displays, extracting information from them can be described in three basic steps.

1. Describe the graph: What does the title say? What is displayed on the x- and y-axis, including the units.
   - Determine the set-up of the graph.
   - Make sure the units used are understood. For example, g·cm^3 means g/cm^3
   - Notice symbols used and check for legend or explanation.

2. Describe the data: Identify the range of data. Are patterns reflected in the data?

3. Interpret the data: How do patterns seen in the graph relate to other things? What conclusions can be drawn from the patterns?

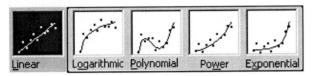

Linear   Logarithmic   Polynomial   Power   Exponential

The relationship being investigated is displayed as the curve of the graph. This curve helps to clarify the relationship: a direct relationship or an inverse relationship for example. Mathematical equations can then be used to express the relationship based on the curve of the graph, d= M/ V or $P_1/T_1 = P_2/T_2$. Values not measured can be predicted by extrapolation or interpolation of the measured values on a graph.

The term best fit curve (or line) is used to describe the smoothest line that can be drawn through the vicinity of the majority of the data points. This is a visual average of the data. Such a curve then describes the relationship between the two variables, the x-axis and the y-axis.

From this curve it is possible to interpolate or extrapolate for additional values that have not been measured. Interpolation is using the curve to read values that lie between data points. Extrapolation involves extending the curve to predict values that lie beyond the data points.

For example, if the graph of the dependent variable vs. the independent variable is a straight line, we say that the dependence is linear. That is the dependent variable depends on the first power of the independent variable. If the independent variable is labeled x and the dependent variable is labeled y and the relationship between y and x is linear, then the relationship can be written in the form y = m x + b, where m and b are constants representing the slope and the y-intercept respectively.

circumference (cm)

diameter (cm)

Connect the dots

Circumference vs diameter of a sphere

circumference (cm)

diameter (cm)

Best fit line

The graph at right shows such a linear relationship. Two points on the graph have been laeled (x1, y1) and (x2, y2).

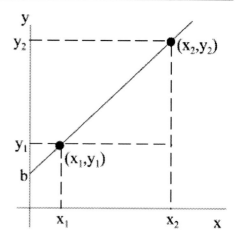

$$slope = \frac{y_2 - y_1}{x_2 - x_1}$$

To find the slope of the graph, select two points near the ends of the **line** and identify the x and y coordinates of each point.

Next find the difference in the y-coordinates ($\Delta y = y_2 - y_1$) and the difference in the x-coordinates ($\Delta x = x_2 - x_1$) between the two points. Don't forget to include units. Then divide Δy by Δx to obtain the slope.

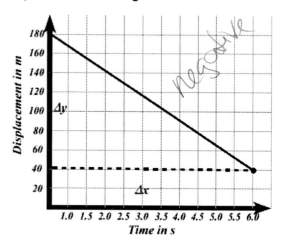

$$slope = \frac{y_2 - y_1}{x_2 - x_1}$$

If the line slopes downward, the slope is negative.

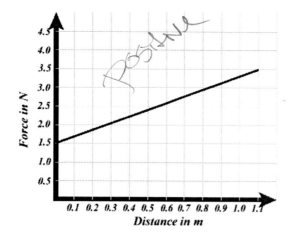

A positive slope value corresponds to an upward sloping line.

Linear graphs show equal changes in variables. However, when the graph of a linear relationship passes through the origin it is a direct proportion. Here, one of the variables is always some multiple of the other.

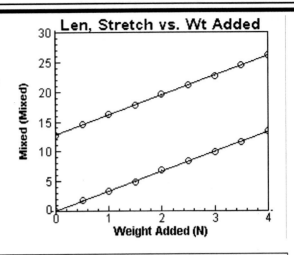

General form: linear relationship	General form direct proportion
y=mx+b	y=mx

Another relationship that can be useful in finding relationships or patterns in data and can be determined from the graph is the inverse proportion. Here, as one variable increases, the other decreases. This relationship can be linear or it can be proportional like in Boyle's Law

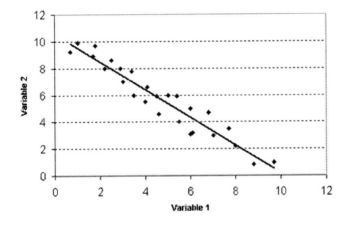

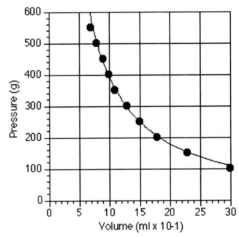

In this case, we say that *y* is inversely proportional to *x*. As *x* increases, *y* decreases. An inverse relationship has an equation of the form,

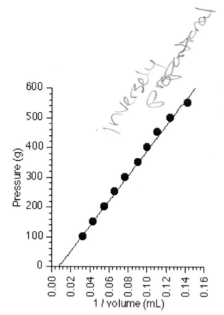

$$y = \frac{C}{x}$$, where *C* is a constant

Graphing the inverse of one of the variables gives a linear graph.

Sometimes the graph obtained from a set of data does not display a linear relationship, the next simplest possibility is a power relationship, i.e. y is directly proportional to some power of x. Mathematically this is written as y = axn, where a is some constant and n is the power mentioned above. A log-log graph can be used to

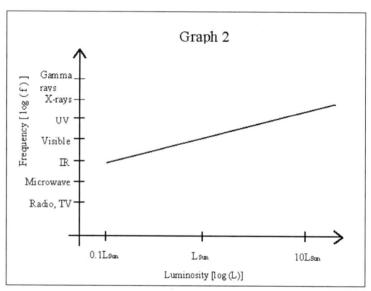

determine whether or not there is a power law relationship between y and x and, if so, what that power is. The constant a can also be determined if desired.

In order to see why the log-log graph is useful, take the log of both sides of the equation above:

y = axn
log y = log (axn)

Now recall that the log of a product is equal to the sum of the logs of the factors in the product, so we have

$$\log y = \log a + \log xn$$
or, rearranging, $\log y = \log xn + \log a$

Now recall the additional property of logs, that $\log xn = n \log x$, so we have

$$\log y = n \log x + \log a$$

Notice that this last equation is in slope intercept form $(y = mx+b)$ if we substitute log y for y and log x for x. Thus if we graph log y on the vertical axis and log x on the horizontal axis, the result will be a straight line. Furthermore the slope of this line will be n and the vertical intercept (i.e. the intercept on the log y axis) will be log a.

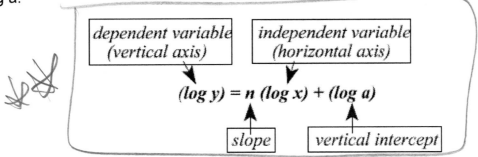

One of the assumptions of science is that ideas are constantly evaluated. In order to achieve this goal, scientists present their findings to a form of peer review either by publishing results or by presenting papers at professional conferences. Hundreds of scientific societies throughout the world publish journals containing articles that report research. The published work is then permanently available to the scientific community. Other scientists are then free to examine the results, repeat the experiments, or take the research further by designing new experiments. Some may critique the original research, pointing out possible errors or alternative interpretations of the data. A scientist may also present research to other scientists at meetings sponsored by one of many scientific societies. In the United States, the American Chemical Society hosts many gatherings of research chemists. It is through presentations like these that the researcher has an opportunity to interact directly with others in the field. As a result of publication and presentation of research, science is a group activity, providing many opportunities for correcting errors and discovering new ideas.

COMPETENCY 0003   UNDERSTAND PRINCIPLES AND PROCEDURES OF MEASUREMENT IN CHEMISTRY.

Dimensional analysis is a structured way to convert units. It involves a conversion factor that allows the units to be cancelled out when multiplied or divided.

These are the steps to converting one dimension measurements.

1. Write the term to be converted, (both number and unit) 6.0 cm = ? km
2. Write the conversion formula(s) 100 cm = 0.00100 km
3. Make a fraction of the conversion formula, such that
   a. if the unit in step 1 is in the numerator, that same unit in step 3 must be in the denominator.
   b. if the unit in step 1 is in the denominator, that same unit in step 3 must be in the numerator.

   Since the numerator and denominator are equal, the fraction must equal 1.

   $$\frac{0.00100 \ km}{100 \ cm} \quad or \quad \frac{100 \ cm}{0.00100 \ km}$$

4. Multiply the term in step 1 by the fraction in step 3. Since the fraction equals 1, you can multiply by it without changing the size of the term.
5. Cancel units : $6.0 \ cm \times \dfrac{0.00100 \ km}{100 \ cm}$

6. Perform the indicated calculation, rounding the answer to the correct number of significant figures.

   $$0.\ 000060 \ km \ \ or \ 6.0 \times 10^{-5} \ km$$

The process is nearly the same for two and three dimension conversions.

Example: How many cm^3 is 1 m^3?

Remember, 100 cm = 1 m and that 1 m^3 is really 1m x 1m x 1m. Substituting in the 100 cm for every meter the problem can be rewritten as 100 cm x 100 cm x 100 cm or $1m^3 = 1\ 000\ 000 \ cm^3$

$1 \ m^3 \times 1\ 000\ 000 \ cm^3/1m^3 = 1\ 000\ 000 \ cm^3$ or $1.0 \times 10^{6} \ cm$.

Example: Convert 4.17 kg/m^2 to g/cm

First to convert from kg to g use 1000 g = 1.00 kg as the conversion factor.

4.17 kg/m^2 x 1000 g/1.00 kg = 4170g /m^2.

Then use 1.00 m = 100 cm to convert the denominator. Remember that m^2 is m x m and replacing m with 100 cm, the denominator becomes 100 cm x 100 cm. or 10 000 cm^2
The conversion factor for the denominator becomes 1.00m^2 = 10 000 cm^2

4170 g/m^2 x 1.00 m^2/10 000 cm^2 = 0.417 g/cm^2

Units are a part of every measurement. Without the units, the numbers could mean many things. For example, the distance 10 could mean 10 cm or 10 m or 10 km. The units are an important part of every measurement. The units will even help solve mathematical problems.

For example: The density of gold is 19.3 g/cm^3. How many grams of gold would be found in 55 cm^3?

Using dimensional analysis, some unit must cancel. The answer needs to be grams so the cm^3 needs to cancel out. Multiply or divide the units so that the cm^3 cancel. In this case

g/cm^3 x cm^3 = g so that is how to solve the problem.

19.3 g/cm^3 x 55 cm^3 = 1060 g of gold.

SI is an abbreviation of the French *Système Internacional d'Unités* or the **International System of Units**. It is the most widely used system of units in the world and is the system used in science. The use of many SI units in the United States is increasing outside of science and technology. There are two types of SI units: **base units** and **derived units**. The base units are:

Quantity	Unit name	Symbol
Length	meter	m
Mass	kilogram	kg
Amount of substance	mole	mol
Time	second	s
Temperature	kelvin	K
Electric current	ampere	A
Luminous intensity	candela	cd

Amperes and candelas are rarely used in chemistry. The name "kilogram" occurs for the SI base unit of mass for historical reasons. Derived units are formed from the kilogram, but appropriate decimal prefixes are attached to the word "gram."

A measurement is **precise** when individual measurements of the same quantity **agree with one another**. A measurement is **accurate** when they **agree with the true value** of the quantity being measured. An **accurate** measurement is **valid**. We get the right answer. A **precise** measurement is **reproducible**. We get a similar answer each time. These terms are related to **sources of error** in a measurement.

Precise measurements are near the **arithmetic mean** of the values. The arithmetic mean is the sum of the measurements divided by the number of measurements. The **mean** is commonly called the **average**. It is the **best estimate** of the quantity.

Random error results from **limitations in equipment or techniques**. A larger **random error decreases precision**. Remember that all measurements reported to proper number of significant digits contain an imprecise final digit to reflect random error.

Systematic error results from **imperfect equipment or technique**. A larger **systematic error decreases accuracy**. Instead of a random error with random fluctuations, there is a result that is too large or small.

Example: An environmental engineering company creates a solution of 5.00 ng/L of a toxin and distributes it to four toxicology labs to test their protocols. Each lab tests the material 5 times. Their results are charted as points on the number lines below. Interpret this data in terms of precision, accuracy, and type of error.

Solution: Results from lab 1 are both accurate and precise when compared to results from the other labs. Results from lab 2 are less precise than those from lab 1. Lab 2 seems to use a protocol that contains a greater random error. However, the mean result from lab 2 is still close to the known value. Lab 3 returned results that were about as precise as lab 1 but inaccurate compared to labs 1 and 2. Lab 3 most likely uses a protocol that yields a systematic error. The data from lab 4 is both imprecise and inaccurate. Systematic and random errors are larger than in lab 1.

Significant figures or **significant digits** are the digits indicating the **precision of a measurement**. There is uncertainty **only** in the last digit.

Example: You measure an object with a ruler marked in millimeters. The reading on the ruler is found to be about 2/3 of the way between 12 and 13 mm. What value should be recorded for its length?

Solution: Recording 13 mm does not give all the information that you found.

Recording $12\frac{2}{3}$ mm implies that an exact ratio was determined.

Recording 12.666 mm gives more information than you found. A value of 12.7 mm or 12.6 mm should be recorded because there is uncertainty only in the last digit.

There are five rules for determining the **number of significant digits** in a quantity.

1) All nonzero digits are significant and all zeros between nonzero digits are significant.

Example: 4.521 kJ and 7002 u both have four significant figures.

2) Zeros to the left of the first nonzero digit are not significant.

Example: 0.0002 m contains one significant figure.

3) Zeros to the right of the decimal point are significant figures.

Example: 32.500 g contains five significant figures.

4) The situation for numbers ending in zeros that are not to the right of the decimal point can be unclear, so **this situation should be avoided** by using scientific notation or a different decimal prefix.
Sometimes a decimal point is used as a placeholder to indicate the units-digit is significant. A word like "thousand" or "million" may be used in informal contexts to indicate the remaining digits are not significant.

Example: 12000 Pa would be considered to have five significant digits by many scientists, but in the context, "The pressure rose from 11000 Pa to 12000 Pa," it almost certainly only has two. "12 thousand pascal" only has two significant figures, but 12000. Pa has five because of the decimal point. The value should be represented as 1.2×10^4 Pa (or 1.2000×10^4 Pa). The best alternative would be to use 12 kPa or 12.000 kPa.

5) Exact numbers have no uncertainty and contain an infinite number of significant figures. These relationships are **definitions**. They are not measurements.

Example: There are exactly 1000 L in one cubic meter.

There are four rules for **rounding off significant figures**.

1) If the leftmost digit to be removed is a four or less, then round down. The last remaining digit stays as it was. Example: Round 43.4 g to 2 significant figures. Answer: 43 g.
2) If the leftmost digit to be removed is a six or more, then round up. The last remaining digit increases by one. Example: Round 6.772 to 2 significant figures. Answer: 6.8 g.
3) If the leftmost digit to be removed is a five that is followed by nonzero digits, then round up. The last remaining digit increases by one. Example: Round 18.502 to 2 significant figures. Answer 19 g.
4) If the leftmost digit to be removed is a five followed by nothing or by only zeros, then force the last remaining digit to be even. If it is odd then round up by increasing it by one. If it is even (including zero) then it stays as it was. Examples: Round 18.50 g and 19.5 g to 2 significant figures. Answers: 18.50 g rounds off to 18 g and 19.5 g rounds off to 20 g.

There are three rules for **calculating with significant figures**.

1) For multiplication or division, the result has the same number of significant figures as the term with the least number of significant figures.

> Example: What is the volume of a compartment in the shape of a rectangular prism 1.2 cm long, 2.4 cm high and 0.9 cm deep?
>
> Solution: Volume=length x height x width.
>
> Volume = 1.2 cm $\times$ 2.4 cm $\times$ 0.9 cm = 2.592 cm (as read on a calculator)
>
> Round to one digit because 0.9 cm has only one significant digit.
>
> Volume = 3 cm^3

2) For addition or subtraction, the result has the same number of digits after the decimal point as the term with the least number of digits after the decimal point.

> Example: Volumes of 250.0 mL, 26 µL, and 4.73 mL are added to a flask. What is the total volume in the flask?
>
> Solution: Only identical units may be added to each other, so 26 µL is first converted to 0.026 mL.
>
> Volume = 250.0 mL + 0.026 mL + 4.73 mL = 254.756 mL (calculator value)
>
> Round to one digit after the decimal because 250.0 mL has only one digit after the decimal. Volume = 254.8 mL.

3) For multi-step calculations, maintain all significant figures when using a calculator or computer and round off the final value to the appropriate number of significant figures after the calculation. When calculating by hand or when **writing down an intermediate value** in a multi-step calculation, maintain the first insignificant digit. In this text, insignificant digits in intermediate calculations are shown in italics except in the examples for the two rules above.

Derived units measure a quantity that may be **expressed in terms of other units**. The derived units important for chemistry are:

Derived quantity	Unit name	Expression in terms of other units	Symbol
Area	square meter	m^2	
Volume	cubic meter	m^3	
	liter	$dm^3 = 10^{-3}\ m^3$	L or l
Mass	unified atomic mass unit	$(6.022 \times 10^{23})^{-1}\ g$	u or Da
Time	minute	60 s	min
	hour	60 min=3600 s	h
	day	24 h=86400 s	d
Speed	meter per second	m/s	
Acceleration	meter per second squared	m/s^2	
Temperature*	degree Celsius	K	°C
Mass density	gram per liter	$g/L = 1\ kg/m^3$	
Amount-of-substance concentration (molarity[†])	molar	mol/L	M
Molality[‡]	molal	mol/kg	*m*
Chemical reaction rate	molar per second[†]	M/s=mol/(L•s)	
Force	newton	$m \cdot kg/s^2$	N
Pressure	pascal	$N/m^2 = kg/(m \cdot s^2)$	Pa
	standard atmosphere[§]	101325 Pa	atm
Energy, Work, Heat	joule	$N \cdot m = m^3 \cdot Pa = m^2 \cdot kg/s^2$	J
	nutritional calorie[§]	4184 J	Cal
Heat (molar)	joule per mole	J/mol	
Heat capacity, entropy	joule per kelvin	J/K	
Heat capacity (molar), entropy (molar)	joule per mole kelvin	J/(mol•K)	
Specific heat	joule per kilogram kelvin	J/(kg•K)	
Power	watt	J/s	W
Electric charge	coulomb	s•A	C
Electric potential, electromotive force	volt	W/A	V
Viscosity	pascal second	Pa•s	
Surface tension	newton per meter	N/m	

*Temperature differences in kelvin are the same as those differences in degrees Celsius. To obtain degrees Celsius from Kelvin, subtract 273.15.
[†]Molarity is considered to be an obsolete unit by some physicists.
[‡]Molality, *m*, is often considered obsolete. Differentiate *m* and meters (m) by context.
[§]These are commonly used non-SI units.

Decimal multiples of SI units are formed by attachi[...]
unit and a symbol prefix directly before the unit sy[...]
10^{-24} to 10^{24}. Only the prefixes you are likely to er[...]
shown below:

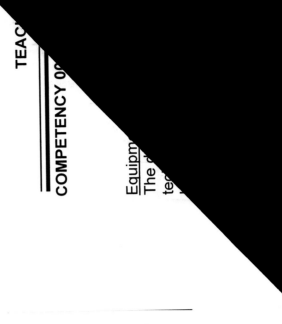

Factor	Prefix	Symbol	Factor
10^9	giga—	G	10^{-1}
10^6	mega—	M	10^{-2}
10^3	kilo—	k	10^{-3}
10^2	hecto—	h	10^{-6}
10^1	deca—	da	10^{-9}
			10^{-12}

Example: 0.0000004355 meters is 4.355×10^{-7} m or 435.5×10^{-9} m. This length is also 435.5 nm or 435.5 nanometers.

Example: Find a unit to express the volume of a cubic crystal that is 0.2 mm on each side so that the number before the unit is between 1 and 1000.

Solution: Volume is length X width X height, so this volume is $(0.0002 \text{ m})^3$ or 8×10^{-12} m^3. Conversions of volumes and areas using powers of units of length must take the power into account. Therefore:

$$1 \text{ m}^3 = 10^3 \text{ dm}^3 = 10^6 \text{ cm}^3 = 10^9 \text{ mm}^3 = 10^{18} \text{ } \mu\text{m}^3 ,$$

The length 0.0002 m is 2×10^2 µm, so the volume is also 8×10^6 µm^3. This volume could also be expressed as 8×10^{-3} mm^3, but none of these numbers are between 1 and 1000.

Expressing volume in liters is helpful in cases like these. There is no power on the unit of liters, therefore:

$$1 \text{ L} = 10^3 \text{ mL} = 10^6 \text{ } \mu\text{L} = 10^9 \text{ nL} .$$

Converting cubic meters to liters gives

$$8 \times 10^{-12} \text{ m}^3 \times \frac{10^3 \text{ L}}{1 \text{ m}^3} = 8 \times 10^{-9} \text{ L} .$$ The crystal's volume is 8 nanoliters (8 nL).

Example: Determine the ideal gas constant, R, in L•atm/(mol•K) from its SI value of 8.3144 J/(mol•K).

Solution: One joule is identical to one m^3•Pa (see the table on the previous page).

$$8.3144 \frac{\text{m}^3 \bullet \text{Pa}}{\text{mol} \bullet \text{K}} \times \frac{1000 \text{ L}}{1 \text{ m}^3} \times \frac{1 \text{ atm}}{101325 \text{ Pa}} = 0.082057 \frac{\text{L} \bullet \text{atm}}{\text{mol} \bullet \text{K}}$$

004 UNDERSTAND PROPER, SAFE, AND LEGAL USE OF EQUIPMENT, MATERIALS, AND CHEMICALS USED IN CHEMISTRY INVESTIGATIONS.

ent
escriptions and diagrams in this skill are included to help you **identify** the chniques. They are **not meant as a guide to perform the techniques** in the lab.

Handling liquids
A **beaker** (below left) is a cylindrical cup with a notch at the top. They are often used for making solutions. An **Erlenmeyer** flask (below center) is a conical flask. A liquid in an Erlenmeyer flask will evaporate more slowly than when it is in a beaker and it is easier to swirl about. A **round-bottom flask** (below-right) is also called a Florence flask. It is designed for uniform heating, but it requires a stand to keep it upright.

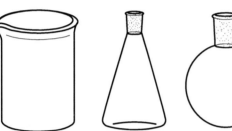

A **test tube** has a rounded bottom and is designed to hold and to heat small volumes of liquid. A **Pasteur pipet** is a small glass tube with a long thin capillary tip and a latex suction bulb.

A **crucible** is a cup-shaped container made of porcelain or metal for holding chemical compounds when heating them to very high temperatures. A **watch glass** is a concave circular piece of glass that is usually used as surface to evaporate a liquid and observe precipitates or crystallization. A **Dewar flask** is a double walled vacuum flask with a metallic coating to provide good thermal insulation and short term storage of liquid nitrogen.

Fitting and cleaning glassware
If a thermometer or funnel must be threaded through a stopper or a piece of tubing and it won't fit, either **make the hole larger or use a smaller piece of glass**. Use soapy water or glycerol to **lubricate** the glass before inserting it. Hold the glass piece as close as possible to the stopper during insertion. It's also good practice to wrap a towel around the glass and the stopper during this time. **Never apply undue pressure**.

Glassware sometimes contains **tapered ground-glass joints** to allow direct glass-to-glass connections. A thin layer of joint **grease** must be applied when assembling an apparatus with ground-glass joints. Too much grease will contaminate the experiment, and too little will permit the components to be permanently locked together. Disassemble the glassware with a **twisting** motion immediately after the experiment is over.

Decimal multiples of SI units are formed by attaching a **prefix** directly before the unit and a symbol prefix directly before the unit symbol. SI prefixes range from 10^{-24} to 10^{24}. Only the prefixes you are likely to encounter in chemistry are shown below:

Factor	Prefix	Symbol	Factor	Prefix	Symbol
10^9	*giga—*	G	10^{-1}	*deci—*	d
10^6	*mega—*	M	10^{-2}	*centi—*	c
10^3	*kilo—*	k	10^{-3}	*milli—*	m
10^2	*hecto—*	h	10^{-6}	*micro—*	μ
10^1	*deca—*	da	10^{-9}	*nano—*	n
			10^{-12}	*pico—*	p

Example: 0.0000004355 meters is 4.355×10^{-7} m or 435.5×10^{-9} m. This length is also 435.5 nm or 435.5 nanometers.

Example: Find a unit to express the volume of a cubic crystal that is 0.2 mm on each side so that the number before the unit is between 1 and 1000.

Solution: Volume is length X width X height, so this volume is $(0.0002\ m)^3$ or $8 \times 10^{-12}\ m^3$. Conversions of volumes and areas using powers of units of length must take the power into account. Therefore:

$$1\ m^3 = 10^3\ dm^3 = 10^6\ cm^3 = 10^9\ mm^3 = 10^{18}\ \mu m^3,$$

The length 0.0002 m is $2 \times 10^2\ \mu m$, so the volume is also $8 \times 10^6\ \mu m^3$. This volume could also be expressed as $8 \times 10^{-3}\ mm^3$, but none of these numbers are between 1 and 1000.

Expressing volume in liters is helpful in cases like these. There is no power on the unit of liters, therefore:

$$1\ L = 10^3\ mL = 10^6\ \mu L = 10^9\ nL.$$

Converting cubic meters to liters gives

$$8 \times 10^{-12}\ m^3 \times \frac{10^3\ L}{1\ m^3} = 8 \times 10^{-9}\ L.$$ The crystal's volume is 8 nanoliters (8 nL).

Example: Determine the ideal gas constant, R, in L•atm/(mol•K) from its SI value of 8.3144 J/(mol•K).

Solution: One joule is identical to one m^3•Pa (see the table on the previous page).

$$8.3144\ \frac{m^3 \bullet Pa}{mol \bullet K} \times \frac{1000\ L}{1\ m^3} \times \frac{1\ atm}{101325\ Pa} = 0.082057\ \frac{L \bullet atm}{mol \bullet K}$$

COMPETENCY 0004   UNDERSTAND PROPER, SAFE, AND LEGAL USE OF EQUIPMENT, MATERIALS, AND CHEMICALS USED IN CHEMISTRY INVESTIGATIONS.

Equipment

The descriptions and diagrams in this skill are included to help you **identify** the techniques. They are **not meant as a guide to perform the techniques** in the lab.

Handling liquids

A **beaker** (below left) is a cylindrical cup with a notch at the top. They are often used for making solutions. An **Erlenmeyer** flask (below center) is a conical flask. A liquid in an Erlenmeyer flask will evaporate more slowly than when it is in a beaker and it is easier to swirl about. A **round-bottom flask** (below-right) is also called a Florence flask. It is designed for uniform heating, but it requires a stand to keep it upright.

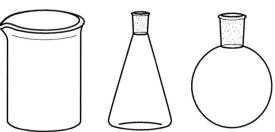

A **test tube** has a rounded bottom and is designed to hold and to heat small volumes of liquid. A **Pasteur pipet** is a small glass tube with a long thin capillary tip and a latex suction bulb.

A **crucible** is a cup-shaped container made of porcelain or metal for holding chemical compounds when heating them to very high temperatures. A **watch glass** is a concave circular piece of glass that is usually used as surface to evaporate a liquid and observe precipitates or crystallization. A **Dewar flask** is a double walled vacuum flask with a metallic coating to provide good thermal insulation and short term storage of liquid nitrogen.

Fitting and cleaning glassware

If a thermometer or funnel must be threaded through a stopper or a piece of tubing and it won't fit, either **make the hole larger or use a smaller piece of glass**. Use soapy water or glycerol to **lubricate** the glass before inserting it. Hold the glass piece as close as possible to the stopper during insertion. It's also good practice to wrap a towel around the glass and the stopper during this time. **Never apply undue pressure**.

Glassware sometimes contains **tapered ground-glass joints** to allow direct glass-to-glass connections. A thin layer of joint **grease** must be applied when assembling an apparatus with ground-glass joints. Too much grease will contaminate the experiment, and too little will permit the components to be permanently locked together. Disassemble the glassware with a **twisting** motion immediately after the experiment is over.

Cleaning glassware becomes more difficult with time, so it should be cleaned soon after the experiment is completed. Wipe off any lubricant with paper towel moistened in a solvent like hexane before washing the glassware. Use a brush with lab soap and water.

Acetone may be used to dissolve most organic residues. Spent solvents should be transferred to a waste container for proper disposal.

Heating

A **hot plate** (shown below) is used to heat Erlenmeyer flasks, beakers and other

containers with a flat bottom. Hot plates often have a built-in **magnetic stirrer**. A **heating mantle** has a hemispherical cavity that is used to heat round-bottom flasks. A **Bunsen burner** is designed to burn natural gas. Burners are useful for heating high-boiling point liquids, water, or solutions of non-flammable materials. They are also used for bending glass tubing. Smooth boiling is achieved by adding **boiling stones** to a liquid.

Boiling and melting point determination

Boiling point is determined by heating the liquid along with a boiling stone in a clamped test tube with a clamped thermometer positioned just above the liquid surface and away from the tube walls. The constant highest-value temperature reading after boiling is achieved is the boiling point.

Melting point is determined by placing pulverized solid in a capillary tube and using a rubber band to fasten the capillary to a thermometer so the sample is at the level of the thermometer bulb. The thermometer and sample are inserted into a **Thiele tube** filled with mineral or silicon oil. The Thiele tube has a sidearm that is heated with a Bunsen burner to create a flow of hot oil. This flow maintains an even temperature during heating. The melting point is read when the sample turns into a liquid. Many **electric melting point devices** are also available that heat the sample more slowly to give more accurate results. These are also safer to use than a Thiele tube with a Bunsen burner.

Centrifugation

A **centrifuge** separates two immiscible phases by spinning the mixture (placed in a **centrifuge tube**) at high speeds. A **microfuge** or microcentrifuge is a small centrifuge. The weight of material placed in a centrifuge must be **balanced**, so if one sample is placed in a centrifuge, a tube with roughly an equal mass of water should be placed opposite the sample.

Filtration

The goal of **gravity filtration** is to remove solids from a liquid and obtain a liquid without solid particulates. Filter paper is folded, placed in a funnel on top of a flask, and wetted with the solvent to seal it to the funnel. Next the mixture is poured through, and the solid-free liquid is collected from the flask.

The goal of **vacuum filtration** is usually to remove liquids from a solid to obtain a solid that is dry. An **aspirator** or a **vacuum pump** is used to provide suction though a rubber tube to a **filter trap**. The trap is attached to a **filter flask** (show to the right) by a second rubber tube. The filter flask is an Erlenmeyer flask with a thick wall and a hose barb for the vacuum tube. Filter flasks are used to filter material using a **Büchner funnel** (shown to the right) or a smaller **Hirsch funnel**. These porcelain or plastic funnels hold a circular piece of filter paper. A single-hole rubber stopper supports the funnel in the flask while maintaining suction.

Mixing

Heterogeneous reaction mixtures in flasks are often mixed by **swirling**. To use a magnetic stirrer, a bar magnet coated with Teflon called a flea or a **stir bar** is slid into the container, and the container is placed on the stirrer. The container should be moved and the stir speed adjusted for smooth mixing. Mechanical stirring paddles, agitators, vortexers, or rockers are also used for mixing.

Decanting

When a course solid has settled at the bottom of a flask of liquid, **decanting** the solution simply means pouring out the liquid and leaving the solid behind.

Extraction

Compounds in solution are often separated based on their **solubility differences**. During **liquid-liquid extraction** (also called **solvent extraction**), a second solvent immiscible to the first is added to the solution in a **separatory funnel** (shown at right). Usually one solvent is nonpolar and the other is a polar solvent like water. The two solvents are immiscible and separate from each other after the mixture is shaken to allow solute exchange. One layer contains the compound of interest, and the other contains impurities to be discarded. The solutions in the two layers are separated from each other by draining liquid through the stopcock.

Titration

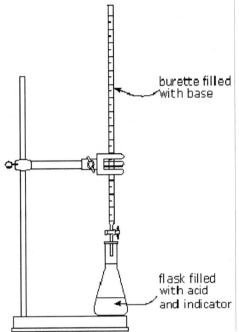

burette filled with base

flask filled with acid and indicator

A titration is a method of analysis that will allow you to determine the precise endpoint of a reaction and therefore the precise quantity of reactant in the titration flask. Titrations can be classified by the chemical reaction involved or by the equipment used to determine the end-point of the titration.

In an acid-base titration a buret is used to deliver the second reactant to the flask and an indicator or pH meter is used to detect the endpoint of the reaction.

In titrimetry, attention is usually focused upon the combination of an ion in the titrant with one of the opposite sign in the titrand solution. Sometimes the combination may involve more than two species, some of which may be nonionic. The combinations may result in precipitation or formation of a complex.

In so-called redox titrations the titrant is usually an oxidizing agent, and is used to determine a substance that can be oxidized and hence can act as a reducing agent.

The passage of a uniform current for a measured period of time can be used to generate a known amount of a product such as a titrant. This fact is the basis of the technique known as **coulometric titration.**

The precision and accuracy with which the end point can be detected is a vital factor in all titrations. Because of its simplicity and versatility, chemical indication is quite common, especially in acid-base titrimetry.

An acid-base indicator is a weak acid or a weak base that changes color when it is transformed from the molecular to the ionized form, or vice versa. The color change is normally intense, so that only a low concentration of indicator is needed.

A potentiometric titration uses a pH meter to determine the end-point of the titration. This method is often employed when a suitable indicator can not be found or high precision is called for.

Conductometric titration is sometimes successful when chemical indication fails.

The underlying principles of conductometric titration are that the solvent and any molecular species in solution exhibit only negligible conductance; that the conductance of a dilute solution rises as the concentration of ions is increased; and that at a given concentration the hydrogen ion and the hydroxyl ion are much better conductors than any of the other ions.

Other common equipment used in titrations include spectrophotometer, microelectrodes. Titrations techniques include thermometric, biamperometric, and nonaquous titration. The proper titration technique is determined by the precision needed and the materials involved in the reaction.

Distillation

Liquids in solution are often separated based on their **boiling point differences**. During simple **distillation**, the solution is placed in a round-bottom flask called the **distillation flask** or **still pot**, and boiling stones are added. The apparatus shown below is assembled (note that clamps and stands are not shown), and the still pot is heated using a heating mantle. Hot vapor during boiling escapes through the **distillation head**, enters the **condenser**, and is cooled and condensed back to a liquid. The vapor loses its heat to water flowing through the outside of the condenser. The condensate or **distillate** falls into the **receiving flask**. The apparatus is open to the atmosphere through a vent above the receiving flask. The distillate will contain a higher concentration of material with the lower boiling point. The less volatile component will achieve a high concentration in the still pot.

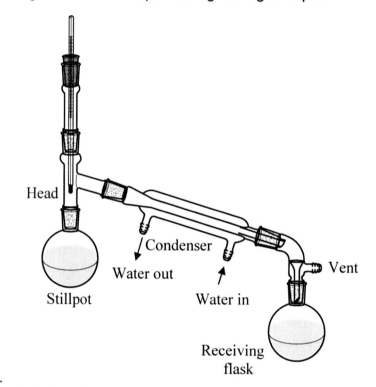

Head temperature is monitored during the process. Distillation may also be used to remove a solid from a pure liquid by boiling and condensing the liquid.

Chemical purchase, use, and disposal

- Inventory all chemicals on hand at least annually. Keep the list up-to-date as chemicals are consumed and replacement chemicals are received.
- If possible, limit the purchase of chemicals to quantities that will be consumed within one year and that are packaged in small containers suitable for direct use in the lab without transfer to other containers.
- Label all chemicals to be stored with date of receipt or preparation and have labels initialed by the person responsible.
- Generally, bottles of chemicals should not remain:
  - Unused on shelves in the lab for more than one week. Move these chemicals to the storeroom or main stockroom.
  - In the storeroom near the lab unused for more than one month. Move these chemicals to the main stockroom.
  - Check shelf life of chemicals. Properly dispose of any out dated chemicals.
- Ensure that the disposal procedures for waste chemicals conform to environmental protection requirements.
- Do not purchase or store large quantities of flammable liquids. Fire department officials can recommend the maximum quantities that may be kept on hand.
- Never open a chemical container until you understand the label and the relevant portions of the MSDS.

Chemical Storage Plan for Laboratories

- Chemicals should be stored according to hazard class (ex. flammables, oxidizers, health hazards/toxins, corrosives, etc.).
- Store chemicals away from direct sunlight or localized heat.
- All chemical containers should be properly labeled, dated upon receipt, and dated upon opening.
- Store hazardous chemicals below shoulder height of the shortest person working in the lab.
- Shelves should be painted or covered with chemical-resistant paint or chemical-resistant coating.
- Shelves should be secure and strong enough to hold chemicals being stored on them. Do not overload shelves.
- Personnel should be aware of the hazards associated with all hazardous materials.
- Separate solids from liquids.

Below are examples of chemical groups that can be used to categorize storage. Use these groups as examples when separating chemicals for compatibility. Please note: reactive chemicals must be more closely analyzed since they have a greater potential for violent reactions.

Acids

- Make sure that all acids are stored by compatibility (ex. separate inorganics from organics).
- Store concentrated acids on lower shelves in chemical-resistant trays or in a corrosives cabinet. This will temporarily contain spills or leaks and protect shelving from residue.
- Separate acids from incompatible materials such as bases, active metals (ex. sodium, magnesium, potassium) and from chemicals which can generate toxic gases when combined (ex. sodium cyanide and iron sulfide).

Bases

- Store bases away from acids.
- Store concentrated bases on lower shelves in chemical-resistant trays or in a corrosives cabinet. This will temporarily contain spills or leaks and protect shelving from residue.

Flammables

- Approved flammable storage cabinets should be used for flammable liquid storage.
- You may store 20 gallons of flammable liquids per 100 sq.ft. in a properly fire separated lab. The maximum allowable quantity for flammable liquid storage in any size lab is not to exceed 120 gallons.
- You may store up to 10 gallons of flammable liquids outside of approved flammable storage cabinets.
- An additional 25 gallons may be stored outside of an approved storage cabinet if it is stored in approved safety cans not to exceed 2 gallons in size.
- Use only explosion-proof or intrinsically safe refrigerators and freezers for storing flammable liquids.

Peroxide-Forming Chemicals

- Peroxide-forming chemicals should be stored in airtight containers in a dark, cool, and dry place.
- Unstable chemicals such as peroxide-formers must always be labeled with date received, date opened, and disposal/expiration date.
- Peroxide-forming chemicals should be properly disposed of before the date of expected peroxide formation (typically 6-12 months after opening).
- Suspicion of peroxide contamination should be immediately investigated. Contact Laboratory Safety for procedures.

Water-Reactive Chemicals

- Water reactive chemicals should be stored in a cool, dry place.
- Do not store water reactive chemicals under sinks or near water baths.
- Class D fire extinguishers for the specific water reactive chemical being stored should be made available.

Oxidizers

- Make sure that all oxidizers are stored by compatibility.
- Store oxidizers away from flammables, combustibles, and reducing agents.

Toxins

- Toxic compounds should be stored according to the nature of the chemical, with appropriate security employed when necessary.
- A "Poison Control Network" telephone number should be posted in the laboratory where toxins are stored. Color coded labeling systems that may be found in your lab:

Hazard	Color Code
Flammables	Red
Health Hazards/Toxins	Blue
Reactives/Oxidizers	Yellow
Contact Hazards	White
General Storage	Gray, Green, Orange

Please Note: Chemicals with labels that are colored and striped may react with other chemicals in the same hazard class. See MSDS for more information. Chemical containers which are not color coded should have hazard information on the label. Read the label carefully and store accordingly.

Schools are regulated by the Environmental Protection Agency, as well as state and local agencies when it comes to disposing of chemical waste.

Check with your state science supervisor, local college or university environmental health and safety specialists and the Laboratory Safety Workshop for advice in the disposal of chemical waste. The American Chemical Society publishes an excellent guidebook, **Laboratory Waste Management, A Guidebook** (1994).

The following are merely guidelines for disposing of chemical waste.

You may dispose of hazardous waste as outlined below. It is the responsibility of the generator to ensure hazardous waste does not end up in ground water, soil or the atmosphere through improper disposal.

1. **Sanitary Sewer** - Some chemicals (acids or bases) may be neutralized and disposed to the sanitary sewer. This disposal option must be approved by the local waste water treatment authority prior to disposal. This may not be an option for some small communities that do not have sufficient treatment capacity at the waste water treatment plant for these types of wastes. Hazardous waste may NOT be disposed of in this manner. This includes heavy metals.

2. **Household Hazardous Waste Facility** - Waste chemicals may be disposed through a county household hazardous waste facility (HHW) or through a county contracted household hazardous waste disposal company. Not all counties have a program to accept waste from schools. Verify with your county HHW facility that they can handle your waste prior to making arrangements.

3. **Disposal Through a Contractor** - A contractor may be used for the disposal of the waste chemicals. Remember that you must keep documentation of your hazardous waste disposal for at least three years. This information must include a waste manifest, reclamation agreement or any written record which describes the waste and how much was disposed, where it was disposed and when it was disposed. Waste analysis records must also be kept when making a determination is necessary. **Any unknown chemicals should be considered hazardous!**

Safety:

Disclaimer: The information presented below is intended as a starting point for identification purposes only and should not be regarded as a comprehensive guide for safety procedures in the laboratory.

It is the responsibility of the readers of this book to consult with professional advisers about safety procedures in their laboratory.

The following list is a summary of the requirements for chemical laboratories contained in the above documents:

1) Dousing shower and eye-wash with a floor drain are required where students handle potentially dangerous materials.
2) Accessible fully-charged fire extinguishers of the appropriate type and fire blankets must be present if a fire hazard exists.
3) There must be a master control valve or switch accessible to and within 15 feet of the instructor's station for emergency cut-off of all gas cocks, compressed air valves, water, or electrical services accessible to students. Valves must completely shut-off with a one-quarter turn. This master control is in addition to the regular main gas supply cut-off, and the main supply cut-off must be shut down upon activation of the fire alarm system.
4) A high capacity emergency exhaust system with a source of positive ventilation must be installed, and signs providing instructions must be permanently installed at the emergency exhaust system fan switch.
5) Fume hoods must contain supply fans that automatically shut down when the emergency exhaust fan is turned on.
6) Rooms and/or cabinets for chemical storage must have limited student access and ventilation to the exterior of the building separate from the air-conditioning system. The rooms should be kept at moderate temperature, be well-illuminated, and contain doors lockable from the outside and operable at all times from the inside. Cabinet shelves must have a half-inch lip on the front and be constructed of non-corrosive material.
7) Appropriate caution signs must be placed at hazardous work and storage areas.

Therefore, all chemistry laboratories should be equipped with the following safety equipment. Both teachers and students should be familiar with the operation of this equipment.

Fire extinguisher: Fire extinguishers are rated for the type of fire it will extinguish. Chemical laboratories should have a combination ABC extinguisher along with a type D fire extinguisher. If a type D extinguisher is not available, a bucket of dry sand will do. Make sure you are trained to use the type of extinguisher available in your setting.

- **Class A** fires are ordinary materials like burning paper, lumber, cardboard, plastics etc.
- **Class B** fires involve flammable or combustible liquids such as gasoline, kerosene, and common organic solvents used in the laboratory.
- **Class C** fires involve energized electrical equipment, such as appliances, switches, panel boxes, power tools, hot plates and stirrers. Water is usually a dangerous extinguishing medium for class C fires because of the risk of electrical shock unless a specialized water mist extinguisher is used.
- **Class D** fires involve combustible metals, such as magnesium, titanium, potassium and sodium as well as pyrophoric organometallic reagents such as alkyllithiums, Grignards and diethylzinc. These materials burn at high temperatures and will react violently with water, air, and/or other chemicals. Handle with care!!
- **Class K** fires are kitchen fires. This class was added to the NFPA portable extinguishers Standard 10 in 1998. Kitche extinguishers installed before June 30, 1998 are "grandfathered" into the standard.

Some fires may be a combination of these! Your fire extinguishers should have ABC ratings on them. These ratings are determined under ANSI/UL Standard 711 and look something like "3-A:40-B:C". Higher numbers mean more firefighting power.

In this example, the extinguisher has a good firefighting capacity for Class A, B and C fires. NFPA has a brief description of UL 711 if you want to know more.

Eyewash:

In the event of an eye injury or chemical splash, use the eyewash immediately.

Help the injured person by holding their eyelids open while rinsing.

Rinse copiously and have the eyes checked by a physician afterwards.

Fire Blanket:

A fire blanket can be used to smother a fire. However, use caution when using a fire blanket on a clothing fire. Some fabrics are polymers that melt onto the skin. Stop, Drop and Roll is the best method for extinguishing clothing on fire.

Safety Shower:

Use a safety shower in the event of a chemical spill. Pull the overhead handle and remove clothing that may be contaminated with chemicals, to allow the skin to be rinsed.

Eye protection:

Everyone present must wear eye protection when anyone in the laboratory is performing any of the following activities:
1) Handling hazardous chemicals
2) Handling laboratory glassware
3) Using an open flame.

Safety glasses do not offer protection from splashing liquids. Safety glasses appear similar to ordinary glasses and may be used in an environment that only requires protection from **flying fragments**. Safety glasses with side-shields offer additional protection from **flying fragments approaching from the side**.

Safety goggles offer protection from both flying fragments and splashing liquids. **Only safety goggles** are suitable for eye protection where **hazardous chemicals** are used and handled.

Safety goggles with no ventilation (type G) or with indirect ventilation (type H) are both acceptable. Goggles should be marked "Z87" to show they meet federal standards.

Skin protection:

Wear gloves made of a material known to resist penetration by the chemical being handled. Check gloves for holes and the absence of interior contamination. Wash hands and arms and clean under fingernails after working in a laboratory.

Wear a lab coat or apron. Wear footwear that completely covers the feet.

Ventilation:

Using a Fume Hood

A fume hood carries away vapors from reagents or reactions you may be working with. Using a fume hood correctly will reduce your personal exposure to potentially harmful fumes or vapors. When using a fume hood, keep the following in mind.

• Place equipment or reactions as far back in the hood as is practical. This will improve the efficiency of fume collection and removal.

• Turn on the light inside the hood using the switch on the outside panel, near the electrical outlets.

• The glass sash of the hood is a safety shield. The sash will fall automatically to the appropriate height for efficient operation and should not be raised above this level, except to move equipment in and out of the hood. Keep the sash between your body and the inside of the hood. If the height of the automatic stop is too high to protect your face and body, lower the sash below this point. Do not stick your head inside a hood or climb inside a hood.

• Wipe up all spills immediately. Clean the glass of your hood, if a splash occurs.

• When you are finished using a hood, lower the sash to the level marked by the sticker on the side.

Work habits

- Never work alone in a laboratory or storage area.
- Never eat, drink, smoke, apply cosmetics, chew gum or tobacco, or store food or beverages in a laboratory environment or storage area.
- Keep containers closed when they are not in use.
- Never pipet by mouth.
- Restrain loose clothing and long hair and remove dangling jewelry.
- Tape all Dewar flasks with fabric-based tape.
- Check all glassware before use. Discard if chips or star cracks are present.
- Never leave heat sources unattended.
- Do not store chemicals and/or apparatus on the lab bench or on the floor or aisles of the lab or storage room.
- Keep lab shelves organized.
- Never place a chemical, not even water, near the edges of a lab bench.
- Use a fume hood that is known to be in operating condition when working with toxic, flammable, and/or volatile substances.
- Never put your head inside a fume hood.
- Never store anything in a fume hood.
- Obtain, read, and be sure you understand the MSDS (see below) for each chemical that is to be used before allowing students to begin an experiment.
- Analyze new lab procedures and student-designed lab procedures in advance to identify any hazardous aspects. Minimize and/or eliminate these components before proceeding. Ask yourself these questions:
  - What are the hazards?
  - What are the worst possible things that could go wrong?
  - How will I deal with them?
  - What are the prudent practices, protective facilities and equipment necessary to minimize the risk of exposure to the hazards?
- Analyze close calls and accidents to eliminate their causes and prevent them from occurring again.
- Identify which chemicals may be disposed of in the drain by consulting the MSDS or the supplier. Clear one chemical down the drain by flushing with water before introducing the next chemical.
- Preplan for emergencies.
  - Keep the fire department informed of your chemical inventory and its location.
  - Consult with a local physician about toxins used in the lab and ensure that your area is prepared in advance to treat victims of toxic exposure.
  - Identify devices that should be shut off if possible in an emergency.
  - Inform your students of the designated escape route and alternate route.

Substitutions

- When feasible, substitute less hazardous chemicals for chemicals with greater hazards in experiments.
- Dilute substances when possible instead of using concentrated solutions.
- Use lesser quantities instead of greater quantities in experiments when possible.
- Use films, videotapes, computer displays, and other methods rather than experiments involving hazardous substances.

Label information

Chemical labels contain safety information in four parts:

1) There will be a signal word. From most to least potentially dangerous, this word will be "Danger!" "Warning!" or "Caution."
2) Statements of hazard (e.g., "Flammable", "May Cause Irritation") follow the signal word. Target organs may be specified.
3) Precautionary measures are listed such as "Keep away from ignition sources" or "Use only with adequate ventilation."
4) First aid information is usually included such as whether to induce vomiting and how to induce vomiting if the chemical is ingested.

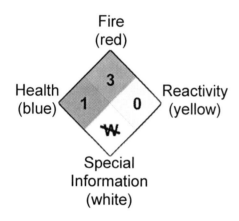

Fire
(red)

Health
(blue)

Reactivity
(yellow)

Special
Information
(white)

Chemical hazard pictorial

Several different pictorials are used on labels to indicate the level of a chemical hazard. The most common is the **"fire diamond" NFPA (National Fire Prevention Association) pictorial** shown at left. A zero indicates a minimal hazard and a four indicates a severe risk. Special information includes if the chemical reacts with water, **OX** for an oxidizer, **COR ACID** for a corrosive acid, and **COR ALK** for a corrosive base. The "Health" hazard level is for **acute toxicity only**.

Pictorials are designed for **quick reference in emergency situations**, but they are also useful as minimal summaries of safety information for a chemical. They are not required on chemicals you purchase, so it's a good idea to add a label pictorial to every chemical you receive if one is not already present. **The entrance to areas where chemicals are stored should carry a fire diamond label** to represent the materials present.

Procedures for flammable materials: minimize fire risk

The vapors of a flammable liquid **or solid** may travel across the room to an ignition source and cause a fire or explosion.

- Store in an approved safety cabinet for flammable liquids. Store in safety cans if possible.
- Minimize volumes and concentrations used in an experiment with flammables.
- Minimize the time containers are open.
- Minimize ignition sources in the laboratory.
- Ensure that there is good air movement in the laboratory before the experiment.
- Check the fire extinguishers and be certain that you know how to use them.
- Tell the students that the "Stop, Drop, and Roll" technique is best for a clothing fire outside the lab, but in the lab they should walk calmly to the safety shower and use it. Practice this procedure with students in drills.
- A fire blanket should not be used for clothing fires because clothes often contain polymers that melt onto the skin. Pressing these fabrics into the skin with a blanket increases burn damage.
- If a demonstration of an exploding gas or vapor is performed, it should be done behind a safety shield using glass vessels taped with fabric tape.

Procedures for corrosive materials: minimize risk of contact

Corrosive materials **destroy or permanently change living tissue** through chemical action. **Irritants** cause inflammation due to an immune response but not through chemical action. The effect is usually reversible but can be severe and long lasting. **Sensitizers** are irritants that cause no symptoms after the first exposure but may cause irritation during a later exposure.

- Always store corrosives below eye level.
- Only diluted corrosives should be used in pre-high school laboratories and their use at full-strength in high school should be limited.
- Dilute corrosive materials by **adding them to water**. Adding water to a concentrated acid or base can cause rapid boiling and splashing.
- Wear goggles and face shield when handling hazardous corrosives. The face, ears, and neck should be protected.
- Wear gloves known to be impervious to the chemical. Wear sleeve gauntlets and a lab apron made of impervious material if splashing is likely.
- Always wash your hands after handling corrosives.
- Splashes on skin should be flushed with flowing water for 15 minutes while a doctor is called.

- If someone splashes a corrosive material on their clothing:
  - First use the safety shower with clothing on.
  - Have them remove all clothing while under the safety shower including shoes and socks. This is no time for modesty.
  - They should stay under the shower for 15 minutes while a doctor is called.
- Splashes in eyes should be dealt with as follows:
  - Take the victim to the eyewash fountain within 30 seconds.
  - Have them hold their eyelids open with thumb and forefinger and use the eyewash.
  - They must continuously move their eyeballs during 15 minutes of rinsing to cleanse the optic nerve at the back of the eye. A doctor should be called.

Procedures for toxic materials: minimize exposure

Toxic effects are either **chronic** or **acute**. Chronic effects are seen after repeated exposures or after one long exposure. Acute effects occur within a few hours at most.

- Use the smallest amount needed at the lowest concentration for the shortest period of time possible. Weigh the risks against the educational benefits.
- Be aware of the five different routes of exposure
  1) Inhalation-the ability to smell a toxin is not a proper indication of unsafe exposure. Work in the fume hood when using toxins. Minimize dusts and mists by cleaning often, cleaning spills rapidly, and maintaining good ventilation in the lab.
  2) Absorption through intact skin-always wear impervious gloves if the MSDS indicates this route of exposure.
  3) Ingestion
  4) Absorption through other body orifices such as ear canal and eye socket.
  5) Injection by a cut from broken contaminated glassware or other sharp equipment.
- Be aware of the first symptoms of overexposure described by the MSDS. Often these are headache, nausea, and dizziness. Get to fresh air and do not return until the symptoms have passed. If the symptom returns when you come back into the lab, contact a physician and have the space tested.
- Be aware of whether vomiting should be induced in case of ingestion
- Be aware of the recommended procedure in case of unconsciousness.

Procedures for reactive materials: minimize incompatibility

Many chemicals are **self-reactive**. For example, they explode when dried out or when disturbed under certain conditions or they react with components of air. These materials generally **should not be allowed into the high school**. Other precautions must be taken to minimize reactions between **incompatible pairs**.

- Store fuels and oxidizers separately.
- Store reducing agents and oxidizing agents separately.
- Store acids and bases separately.
- Store chemicals that react with fire-fighting materials (i.e., water or carbon dioxide) under conditions that minimize the possibility of a reaction if a fire is being fought in the storage area.
- MSDSs list other incompatible pairs.
- Never store chemicals in alphabetical order by name.
- When incompatible pairs must be supplied to students, do so under direct supervision with very dilute solutions and/or small quantities.

Material Safety Data Sheet (MSDS) information

Many chemicals have a mixture of toxic, corrosive, flammability, and reactivity risks. The **Material Safety Data Sheet** or **MSDS** for a chemical contains detailed safety information that is not presented on the label. This includes acute and chronic health effects, first aid and firefighting measures, what to do in case of a spill, and ecological and disposal considerations.

The MSDS will state whether the chemical is a known or suspected carcinogen, mutagen, or teratogen. A **carcinogen** is a compound that causes cancer (malignant tumors). A **mutagen** alters DNA with the potential effects of causing cancer or birth defects in unconceived children. A **teratogen** produces birth defects and acts during fetal development

There are many parts of an MSDS that are not written for the layperson. Their level of detail and technical content intimidate many people outside the fields of toxicology and industrial safety. According to the American Chemical Society (http://membership.acs.org/c/ccs/pubs/chemical_safety_manual.pdf), an MSDS places "an over-emphasis on the toxic characteristics of the subject chemical."

In a high school chemistry lab, the value of an MSDS is in the words and not in the numerical data it contains, but some knowledge of the numbers is useful for comparing the dangers of one chemical to another. Numerical results of animal toxicity studies are often presented in the form of LD_{50} **values**. These represent the **dose required to kill 50% of animals** tested.

Exposure limits may be presented in three ways:

1) PEL (Permissible Exposure Limit) or TLV-TWA (Threshold Limit Value-Time Weighted Average). This is the maximum permitted concentration of the airborne chemical in volume parts per million (ppm) for a **worker exposed 8 hours daily**.
2) TLV-STEL (Threshold Limit Value-Short Term Exposure Limit). This is the maximum concentration permitted for a 15-minute exposure period.
3) TLV-C (Threshold Limit Value-Ceiling). This is the concentration that should never be exceeded at any moment.

http://hazard.com/msds/index.php contains a large database of MSDSs. http://www.ilpi.com/msds/ref/demystify.html contains a useful "MSDS demystifier." Cut and paste an MSDS into the web page, and hypertext links will appear to a glossary of terms.

Facilities and equipment

- Use separate labeled containers for general trash, broken glass, for each type of hazardous chemical waste—ignitable, corrosive, reactive, and toxic.
- Keep the floor area around safety showers, eyewash fountains, and fire extinguishers clear of all obstructions.
- Never block escape routes.
- Never prop open a fire door.
- Provide safety guards for all moving belts and pulleys.
- Instruct everyone in the lab on the proper use of the safety shower and eyewash fountain (see Corrosive materials above). Most portable eyewash devices cannot maintain the required flow for 15 minutes. A permanent eyewash fountain is preferred.
- If contamination is suspected in the breathing air, arrange for sampling to take place.
- Regularly inspect fire blankets, if present, for rips and holes. Maintain a record of inspection.
- Regularly check safety showers and eyewash fountains for proper rate of flow. Maintain a record of inspection.
- Keep up-to-date emergency phone numbers posted next to the telephone.
- Place fire extinguishers near an escape route.
- Regularly maintain fire extinguishers and maintain a record of inspection. Arrange with the local fire department for training of teachers and administrators in the proper use of extinguishers.
- Regularly check fume hoods for proper airflow. Ensure that fume hood exhaust is not drawn back into the intake for general building ventilation.
- Secure compressed gas cylinders at all times and transport them only while secured on a hand truck.
- Restrict the use and handling of compressed gas to those who have received formal training.

- Install chemical storage shelves with lips. Never use stacked boxes for storage instead of shelves.
- Only use an explosion-proof refrigerator for chemical storage.
- Have appropriate equipment and materials available in advance for spill control and cleanup. Consult the MSDS for each chemical to determine what is required. Replace these materials when they become outdated.
- Provide an appropriate supply of first aid equipment and instruction on its proper use.

Additional comments: Teach safety to students

- Weigh the risks and benefits inherent in lab work, inform students of the hazards and precautions involved in their assignment, and involve students in discussions about safety before every assignment.
- If an incident happens, it can be used to improve lab safety via student participation. Ask the student involved. The student's own words about what occurred should be included in the report.
- Safety information supplied by the manufacturer on a chemical container should be seen by students who actually use the chemical. If you distribute chemicals into smaller containers to be used by students, copy the hazard and precautionary information from the original label onto the labels for the students' containers. Students interested in graphic design may be able to help you perform this task. Labels for many common chemicals may be found here: http://beta.ehs.cornell.edu/labels/cgi-bin/label_selection.pl
- Organize a student safety committee whose task is to conduct one safety inspection and present a report. A different committee may be organized each month or every other month.

Additional comments: General

- Every chemical is hazardous. The way it is used determines the probability of harm.
- Every person is individually and personally responsible for the safe use of chemicals.
- If an accident might happen, it will eventually happen. Proper precautions will ensure the consequences are minimized when it does occur.
- Every accident is predicted by one or more **close calls** where nobody is injured and no property is damaged but something out of the ordinary occurred. Examples might be a student briefly touching a hot surface and saying "Ouch!" with no injury, two students engaged in horseplay, or a student briefly removing safety goggles to read a meniscus level. **Eliminate the cause of a close call and you have stopped a future accident.**

Also see:
http://www.labsafety.org/40steps.htm,
http://www.flinnsci.com/Sections/Safety/safety.asp,
and the American Chemical Society safety publications listed under References.

It was as if you fired a 15-inch shell at a sheet of tissue paper and it came back to hit you. — Ernest Rutherford (after scattering alpha particles off gold foil)

COMPETENCY 0005   UNDERSTAND THE CONCEPT OF MATTER, AND ANALYZE CHEMICAL AND PHYSICAL PROPERTIES OF AND CHANGES IN MATTER.

The word "matter" describes everything that has physical existence, i.e. has mass and takes up space. However, the make up of matter allows it to be separated into categories. The two main classes of matter are **pure substance and mixture.** Each of these classes can also be divided into smaller categories such as element, compound, homogeneous mixture or heterogeneous mixture based on composition.

PURE SUBSTANCES: A pure substance is a form of matter with a definite composition and distinct properties. This type of matter can not be separated by ordinary processes like filtering, centrifuging, boiling or melting.

Pure substances are divided into elements and compounds.

Elements: A single type of matter, called an atom, is present. Elements can not be broken down any farther by ordinary chemical processes. They are the smallest whole part of a substance that still represents that substance.

Compounds: Two or more elements chemically combined are present. A compound may be broken down into its elements by chemical processes such as heating or electric current. Compounds have a uniform composition regardless of the sample size or source of the sample.

MIXTURES: Two or more pure substances that are not chemically combined are present. Mixtures may be of any proportion and can be physically separated by processes like filtering, centrifuging, boiling or melting.

Mixtures can be classified according to particle size.

Homogeneous Mixtures: Homogeneous mixtures have the same composition and properties throughout the mixture and are also known as solutions. They have a uniform color and distribution of solute and solvent particles throughout the mixture.

Heterogeneous Mixtures: Heterogeneous mixtures do not have a uniform distribution of particles throughout the mixture. The different components of the mixture can easily be identified and separated.

A mixture consists of two or more substances that when put together retain their individual physical and chemical properties. That is, no new substances are formed. Differences in physical and chemical properties of the components of the mixture can be used to separate the mixture. For example, salt and pepper have different colors and can be separated by physically moving the white crystals away from the dark particles. Or salt dissolves in water while pepper does not and the two can be separated by adding water to the mixture. The salt will dissolve while the pepper will float on top of the water. Skim off the pepper and then evaporate the water. The salt and pepper are now separated.

Common methods for separating mixtures include filtration, chromatography, distillation and extraction. **Filtering** relies on one of the substances being soluble in the solvent, and the other being insoluble. The soluble substance ends up in the **filtrate** (the liquid that passes through the filter paper). The insoluble substance ends up in the filter paper as the **residue**.

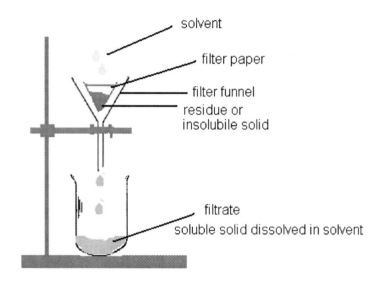

solvent

filter paper

filter funnel

residue or insolubile solid

filtrate
soluble solid dissolved in solvent

Distillation and **fractional distillation** rely on the fact that the liquids in the mixture are different. Each liquid has its own unique **boiling point**. As the temperature rises, it reaches the lowest boiling point liquid and so this liquid boils. The temperature remains at that temperature until all of the liquid has boiled away. The vapor passes into a condenser and is cooled and so turns back

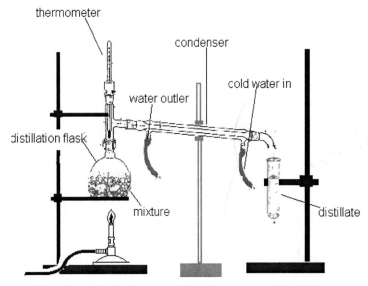

into a liquid. The pure liquid is now in a different part of the apparatus and rolls down into a test tube. This liquid has been separated from the mixture.

Chromatography is a technique, which is used to separate mixtures based on differential migration. Differential migration uses differences in physical properties such as solubility, molecular size and polarity to separate the substance. The mixture is placed in a solution, which becomes the mobile phase. This solution then passes over a stationary phase. Different components of the mixture travel at different rates. After the separation, the time that component took to emerge from the instrument (or its location within the stationary phase) is found with a detector. These chromatographic separations have many different types of stationary phases including paper, silica on glass or plastic plates (thin-layer chromatography), volatile gases (gas chromatography) and liquids, (liquid chromatography including column and high-performance liquid chromatography or HPLC) as well as gels (electrophoresis).

The result is a **chromatogram** like the one shown below. The **identity of an unknown** peak is found by comparing its location on the chromatogram to standards. The **concentration** of a component is found from signal strength or peak area by comparison to calibration curves of known concentrations.

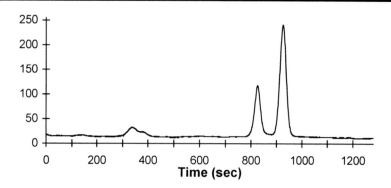

In **paper chromatography** and **thin layer chromatography (TLC)**, the sample rises up by capillary action through a solid phase.

In **gas chromatography (GC)**, the sample is vaporized and forced through a column filled with a packing material. GC is often used to separate and determine the concentration of **low molecular weight volatile organic** compounds.

In **liquid chromatography (LC)**, the sample is a liquid. It is either allowed to seep through an open column using the force of gravity or it is forced through a closed column under pressure. The variations of liquid chromatography depend on the identity of the packing material. For example, an ion-exchange liquid chromatograph contains a material with a charged surface. The mixture components with the opposite charge interact with this packing material and spend more time in the column. LC is often used to separate **large organic polymers** like proteins.

Each component in the mixture can be identified by its retention factor or retention time once it has been separated. Both retention factor, R_f, and retention time are quantitative indications of how far a particular compound travels in a particular solvent or how long the component is retained on the column before eluting with the mobile phase

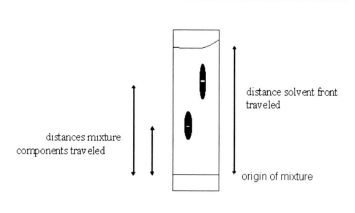

(for gas chromatography and liquid chromatography). By comparing the R_f of the components in the mixture to known values for substances, identification of the components in the mixture can be achieved.

R_f = distance the component color traveled from the point of application
distance the solvent front traveled from the point of application

In an ideal chemical world, reactions would produce products in exactly the right amount and in exactly the right way; no product loss, no competing side reactions. But then things would be very simple for the chemist. In the real chemical world, things are not so simple for the chemist. Most chemical reactions produce product intermingled with by-products or impurities.

Extraction is a method used to separate the desired product from impurities. This process is used in product purification and results from an unequal distribution of solute between two immiscible solvents. This process, then, makes use of differences is solubilities to separate components in what is called a work-up. The work-up is a planned sequence of extracting and washing.

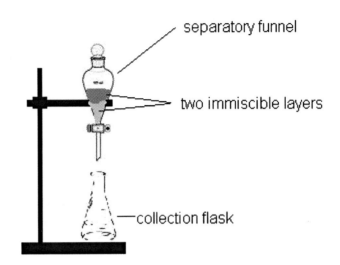

separatory funnel

two immiscible layers

collection flask

The planning involves determining the two solvents. One is usually water, the aqueous layer, while the other is an organic solvent, the organic layer. The important feature here is that the two solvents are immiscible and the mixture is the partition between the two solvents. The mixture is then separated as its components are attracted to the solvents differently. Most charged particles and inorganic salts will move towards the aqueous layer and dissolve in it, while the neutral organic molecules will move towards the organic layer

A **physical property** of matter is a property that can be determined without inducing a chemical change. All matter has mass and takes up space with an associated size. Matter experiencing gravity has a weight. Most matter we encounter exists in one of three phases. Some other examples of physical properties of matter include.

Melting point refers to the temperature at which a solid becomes a liquid. **Boiling point** refers to the temperature at which a liquid becomes a gas. Melting takes place when there is sufficient energy available to break the intermolecular forces that hold molecules together in a solid. Boiling occurs when there is enough energy available to break the intermolecular forces holding molecules together as a liquid.

Hardness describes how difficult it is to scratch or indent a substance. The hardest natural substance is diamond.

Density measures the mass of a unit volume of material. Units of g/cm^3 are commonly used. SI base units of kg/m^3 are also used. One g/cm^3 is equal to one thousand kg/m^3. Density (ρ) is calculated from mass (m) and volume (V) using the formula:

$$\rho = \frac{m}{V}.$$

The above expression is often manipulated to determine the mass of a substance if its volume and density are known ($m = \rho V$) or the volume of a substance if its mass and density are known ($V = m / \rho$).

Electrical conductivity measures a material's ability to conduct an electric current. The high conductivity of metals is due to the presence of metallic bonds. The high conductivity of electrolyte solutions is due to the presence of ions in solution.

Chemical characteristics or behavior are the ways substances interact with each other or the substance's ability to form different substances. Chemical properties are observable only during a chemical reaction.

Some common chemical properties are

Flammability: some substances have the property to react with oxygen under elevated temperature

Acidity: some substances have the property to produce hydrogen ions when in an aqueous solution.

Oxidation-reduction: some substances have the ability to gain or lose electrons. Some examples are tarnishing, rusting, patination.

Precipitation: some substances have the ability to combine with others, changing their solubility.

A **physical change** does not create a new substance. **Atoms are not rearranged into different compounds**. The material has the same chemical composition as it had before the change. Changes of state as described in the previous section are physical changes. Frozen water or gaseous water is still H_2O. Taking a piece of paper and tearing it up is a physical change. You simply have smaller pieces of paper.

Iron Nail   Bent   Iron Nail:Physical change—still an iron nail

A **chemical change** is a chemical reaction. It **converts one substance into another** because atoms are rearranged to form a different compound. Paper undergoes a chemical change when you burn it. You no longer have paper. A chemical change to a pure substance alters its properties.

 Iron Nail   Rusty Iron Nail: Chemical Change---iron oxide (rust) is present

COMPETENCY 0006   UNDERSTAND THE VARIOUS MODELS OF ATOMIC STRUCTURE, THE PRINCIPLES OF QUANTUM THEORY, AND THE PROPERTIES AND INTERACTIONS OF SUBATOMIC PARTICLES.

Ancient philosophers in Greece, India, China, and Japan speculated that all matter was composed of four or five elements. The Greeks thought that these were: fire, air, earth, and water. Indian philosophers and the Greek Aristotle also thought a fifth element—"aether" or "quintessence"—filled all of empty space.

However, some disagreement could be found. Democritus believed that there existed a piece of matter so small that it could not be divided and still retain properties of the matter. He called this piece of matter an atom. However, it took almost two thousand years for there to be any experimental evidence to support the existence of atoms. The alchemists laid the foundation for an experimental science and early scientists like Lavoisier, Preistly, Boyle, Avogadro and Proust along with Newton developed ideas on the composition of matter it wasn't until the early 1800's that these ideas were acted on.

Dalton's table of atomic symbols and masses

Dalton
The existence of fundamental units of matter called atoms of different types called elements was proposed by ancient philosophers without any evidence to support the belief. Modern atomic theory is credited to the work of **John Dalton** published in 1803-1807.

Observations made by him and others about the composition, properties, and reactions of many compounds led him to develop the following postulates:

1) Each element is composed of small particles called atoms.
2) All atoms of a given element are identical in mass and other properties.
3) Atoms of different elements have different masses and differ in other properties.
4) Atoms of an element are not created, destroyed, or changed into a different type of atom by chemical reactions.
5) Compounds form when atoms of more than one element combine.
6) In a given compound, the relative number and kind of atoms are constant.

Dalton determined and published the known relative masses of a number of different atoms. He also formulated the law of partial pressures. Dalton's work focused on the ability of atoms to arrange themselves into molecules and to rearrange themselves via chemical reactions, but he did not investigate the composition of atoms themselves. **Dalton's model of the atom** was a tiny, indivisible, indestructible **particle** of a certain mass, size, and chemical behavior, but Dalton did not deny the possibility that atoms might have a substructure.

Prior to the late 1800s, atoms, following Dalton's ideas, were thought to be small, spherical and indivisible particles that made up matter. However, with the discovery of electricity and the investigations that followed, this view of the atom changed.

Thomson
Joseph John Thomson, often known as **J. J. Thomson**, was the first to examine this substructure. In the mid-1800s, scientists had studied a form of radiation called "cathode rays" or "electrons" that originated from the negative electrode (cathode) when electrical current was forced through an evacuated tube. Thomson determined in 1897 that **electrons have mass**, and because many different cathode materials release electrons, Thomson proposed that the **electron is a subatomic particle**. **Thomson's model of the atom** was a uniformly positive particle with electrons contained in the interior. This has been called the "plum-pudding" model of the atom where the pudding represents the uniform sphere of positive electricity and the bits of plum represent electrons. For more on Thomson, see http://www.aip.org/history/electron/jjhome.htm.

Planck
Max Planck determined in 1900 that **energy is transferred by radiation in exact multiples of a discrete unit of energy called a quantum**. Quanta of energy are extremely small, and may be found from the frequency of the radiation, v, using the equation:

$$\Delta E = hv$$

where h is Planck's constant and hv is a quantum of energy.

Rutherford

Ernest Rutherford studied atomic structure in 1910-1911 by firing a beam of alpha particles at thin layers of gold leaf. According to Thomson's model, the path of an alpha particle should be deflected only slightly if it struck an atom, but Rutherford observed some alpha particles bouncing almost backwards, suggesting that **nearly all the mass of an atom is contained in a small positively charged nucleus**. **Rutherford's model of the atom** was an analogy to the sun and the planets. A small positively charged nucleus is surrounded by circling electrons and mostly by empty space. Rutherford's experiment is explained in greater detail in this flash animation: http://www.mhhe.com/physsci/chemistry/essentialchemistry/flash/ruther14.swf.

Bohr

Niels Bohr incorporated Planck's quantum concept into Rutherford's model of the atom in 1913 to explain the **discrete frequencies of radiation emitted and absorbed by atoms with one electron** (H, He^+, and Li^{2+}). This electron is attracted to the positive nucleus and is closest to the nucleus at the **ground state** of the atom. When the electron absorbs energy, it moves into an orbit further from the nucleus and the atom is said to be in an electronically **excited state**. If sufficient energy is absorbed, the electron separates from the nucleus entirely, and the atom is ionized:

$$H \rightarrow H^+ + e^-$$

The energy required for ionization from the ground state is called the atom's **ionization energy**. The discrete frequencies of radiation emitted and absorbed by the atom correspond (using Planck's constant) to discrete energies and in turn to discrete distances from the nucleus. **Bohr's model of the atom** was a small positively charged nucleus surrounded mostly by empty space and by electrons orbiting at certain discrete distances ("shells") corresponding to discrete energy levels. Animations utilizing the Bohr model may be found at the following two URLs: http://artsci-ccwin.concordia.ca/facstaff/a–c/bird/c241/D1.html and http://www.mhhe.com/physsci/chemistry/essentialchemistry/flash/linesp16.swf.

Bohr's model of the atom didn't quite fit experimental observations for atoms other than hydrogen. He was, however, on the right track. DeBroglie was the first to suggest that possibly matter behaved like a wave. Until then, waves had wave properties of wavelength, frequency and amplitude while matter had matter properties like mass and volume. DeBroglie's suggestion was quite unique and interesting to scientists.

De Broglie

Depending on the experiment, radiation appears to have wave-like or particle-like traits. In 1923-1924, Louis de Broglie applied this **wave/particle duality to all matter with momentum**. The discrete distances from the nucleus described by Bohr corresponded to permissible distances where standing waves could exist. **De Broglie's model of the atom** described electrons as **matter waves in standing wave orbits** around the nucleus. The first three standing waves corresponding to the first three discrete distances are shown in the figure. De Broglie's model may be found here: http://artsci-ccwin.concordia.ca/facstaff/a-c/bird/c241/D1-part2.html.

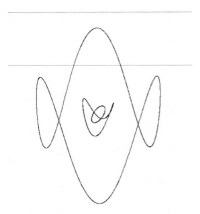

Heisenberg

The realization that both matter and radiation interact as waves led Werner Heisenberg to the conclusion in 1927 that the act of observation and measurement requires the interaction of one wave with another, resulting in an **inherent uncertainty** in the location and momentum of particles. This inability to measure phenomena at the subatomic level is known as the **Heisenberg uncertainty principle**, and it applies to the location and momentum of electrons in an atom. A discussion of the principle and Heisenberg's other contributions to quantum theory is located here: http://www.aip.org/history/heisenberg/.

Schrödinger

When Erwin Schrödinger studied the atom in 1925, he replaced the idea of precise orbits with regions in space called **orbitals** where electrons were likely to be found. **The Schrödinger equation** describes the **probability** that an electron will be in a given region of space, a quantity known as **electron density** or Ψ^2. The diagrams below are surfaces of constant Ψ^2 found by solving the Schrödinger equation for the hydrogen atom $1s$, $2p_z$ and $3d_0$ orbitals. Additional representations of solutions may be found here: http://library.wolfram.com/webMathematica/Physics/Hydrogen.jsp.
Schrödinger's model of the atom is a mathematical formulation of quantum mechanics that describes the electron density of orbitals. It is the atomic model that has been in use from shortly after it was introduced up to the pre

$1s$

$2p_z$

$3d_z{}^2$

This model explains the movement of electrons to higher energy levels when exposed to energy. It also explains the movement of electrons to lower energy levels when the source of energy has disappeared. Accompanying this drop in energy level is the emission of electromagnetic radiation (light as one possibility).

Using Plank's equation, the frequency of that electromagnetic radiation, EMR, can be determined, radiation caused by the movement of matter.

Quantum numbers

The quantum-mechanical solutions from the Schrödinger Equation utilize three quantum numbers (n, l, and m_l) to describe an orbital and a fourth (m_s) to describe an electron in an orbital. This model is useful for understanding the frequencies of radiation emitted and absorbed by atoms and chemical properties of atoms.

The **principal quantum number n** may have positive integer values (1, 2, 3, …). n is a measure of the **distance** of an orbital from the nucleus, and orbitals with the same value of n are said to be in the same **shell**. This is analogous to the Bohr model of the atom. Each shell may contain up to $2n^2$ electrons.

The **azimuthal quantum number l** may have integer values from 0 to n-1. l describes the angular momentum of an orbital. This determines the orbital's **shape**. Orbitals with the same value of n and l are in the same **subshell**, and each subshell may contain up to ($4l + 2$ electrons). Subshells are usually referred to by the principle quantum number followed by a letter corresponding to l as shown in the following table:

Azimuthal quantum number l	0	1	2	3	4
Subshell designation	s	p	d	f	g

The **magnetic quantum number m_l or m** may have integer values from $-l$ to l. m_l is a measure of how an individual orbital responds to an external magnetic field, and it often describes an orbital's **orientation**. A subscript—either the value of m_l or a function of the x-, y-, and z-axes—is used to designate a specific orbital. Each orbital may hold up to two electrons.

The **spin quantum number m_s or s** has one of two possible values: $-1/2$ or $+1/2$. m_s differentiates between the two possible electrons occupying an orbital. Electrons moving through a magnet behave as if they were tiny magnets themselves spinning on their axis in either a clockwise or counterclockwise direction. These two spins may be described as $m_s = -1/2$ and $+1/2$ or as down and up.

The **Pauli exclusion principle** states that **no two electrons in an atom may have the same set of four quantum numbers.**

The following table summarizes the relationship among n, l, and m_l through $n=3$:

n	l	Subshell	m_l	Orbitals in subshell	Maximum number of electrons in subshell
1	0	1s	0	1	2
2	0	2s	0	1	2
	1	2p	−1, 0, 1	3	6
3	0	3s	0	1	2
	1	3p	−1, 0, 1	3	6
	2	3d	−2, −1, 0, 1, 2	5	10

Subshell energy levels

In single-electron atoms (H, He$^+$, and Li^{2+}) above the ground state, subshells within a shell are all at the same energy level, and an orbital's energy level is only determined by n. However, in all other atoms, multiple electrons repel each other. Electrons in orbitals closer to the nucleus create a screening or **shielding effect** on electrons further away from the nucleus, preventing them from receiving the full attractive force of the nucleus. **In multi-electron atoms, both n and l determine the energy level of an orbital**. In the absence of a magnetic field, **orbitals in the same subshell with different m_l all have the same energy** and are said to be **degenerate orbitals**.

The following list orders subshells by increasing energy level:
$1s < 2s < 2p < 3s < 3p < 4s < 3d < 4p < 5s < 4d < 5p < 6s < 4f < 5d < 6p < 7s < 5f < ...$

This list may be constructed by arranging the subshells according to n and l and drawing diagonal arrows as shown below:

1s

2s 2p

3s 3p 3d

4s 4p 4d 4f

5s 5p 5d 5f 5g

6s 6p 6d 6f 6g

7s 7p 7d 7f 7g

8s 8p 8d 8f 8g

Electron arrangements (also called electron shell structures) in an atom may be represented using three methods: an **electron configuration**, an **orbital diagram**, or an **energy level diagram**.

All three methods require knowledge of the subshells occupied by electrons in a certain atom. The **Aufbau principle** or **building-up rule** states that **electrons at ground state fill orbitals starting at the lowest available energy levels**.

An **electron configuration** is a **list of subshells** with superscripts representing the **number of electrons** in each subshell.

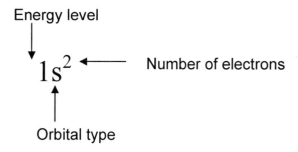

Energy level

$1s^2$ ← Number of electrons

Orbital type

For example, an atom of boron has 5 electrons. According to the Aufbau principle, two will fill the 1s subshell, two will fill the higher energy 2s subshell, and one will occupy the 2p subshell which has an even higher energy. The electron configuration of boron is $1s^2 2s^2 2p^1$. Similarly, the electron configuration of a vanadium atom with 23 electrons is:

$$1s^2 2s^2 2p^6 3s^2 3p^6 4s^2 3d^3.$$

Configurations are also written with their principle quantum numbers together:

$$1s^2 2s^2 2p^6 3s^2 3p^6 3d^3 4s^2.$$

Electron configurations are often written to emphasize the outermost electrons. This is done by writing the symbol in brackets for the element with a full p subshell from the previous shell and adding the **outer electron configuration** onto that configuration. The element with the last full p subshell will always be a noble gas from the right-most column of the periodic table. For the vanadium example, the element with the last full p subshell has the configuration $1s^2 2s^2 2p^6 3s^2 3p^6$. This is $_{18}$Ar. The configuration of vanadium may then be written as $[Ar]4s^2 3d^3$ where $4s^2 3d^3$ is the outer electron configuration.

Example: Write the electron configuration for Bromine (35 electrons):

Solution:

Using the diagram above

Fill the orbitals in order with the required electrons until the total electrons (superscript) is 35.

$1s^2\ 2s^2\ 2p^6\ 3s^2\ 3p^6\ 4s^2\ 3\,d^{10}\ 4p^5$

Orbital diagrams assign electrons to individual orbitals so the energy state of individual electrons may be found. This requires knowledge of how electrons occupy orbitals within a subshell. **Hund's rule** states that **before any two electrons occupy the same orbital, other orbitals in that subshell must first contain one electron each with parallel spins**. Electrons with up and down spins are shown by half-arrows, and these are placed in lines of orbitals (represented as boxes or dashes) according to Hund's rule, the Aufbau principle, and the Pauli exclusion principle. Below is the orbital diagram for vanadium:

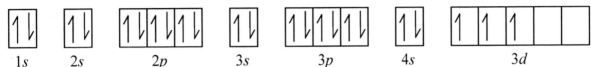

An **energy level diagram** is an orbital diagram that shows subshells with higher energy levels higher up on the page. The energy level diagram of vanadium is:

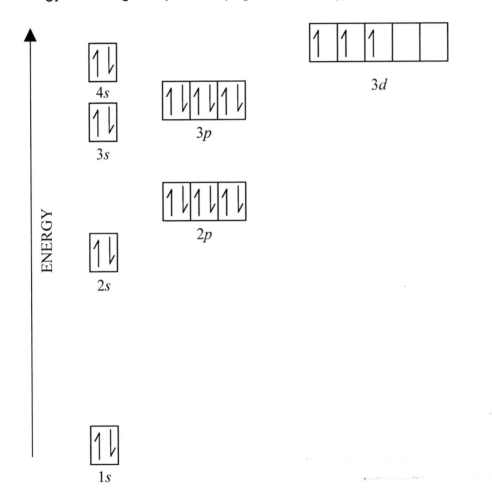

COMPETENCY 0007 UNDERSTAND THE ORGANIZATION OF THE PERIODIC TABLE.

The first periodic table was developed in 1869 by Dmitri Mendeleev several decades before the nature of electron energy states in the atom was known. Mendeleev arranged the elements in order of increasing atomic mass into **columns of similar physical and chemical properties**. He then boldly **predicted the existence and the properties of undiscovered elements** to fill the gaps in his table. These interpolations were initially treated with skepticism until three of Mendeleev's theoretical elements were discovered and were found to have the properties he predicted. It is the correlation with properties—not with electron arrangements—that have placed the periodic table at the beginning of most chemistry texts.

The elements in a column are known as a group, and groups are numbered from 1 to 18. Older numbering styles used roman numerals and letters. **A row of the periodic table is known as a period**, and periods of the known elements are numbered from 1 to 7. The lanthanoids are all in period 6, and the actinoids are all in period 7.

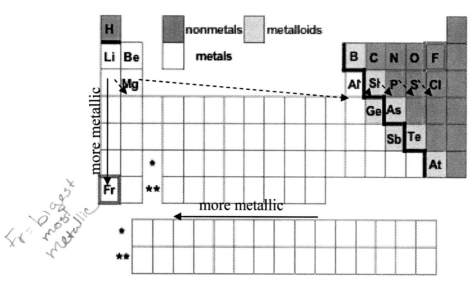

Metals, nonmetals, and atomic radius

Elements in the periodic table are divided into the two broad categories of **metals** and **nonmetals** with a jagged line separating the two as shown in the figure.

Seven elements near the line dividing metals from nonmetals exhibit some properties of each and are called **metalloids** or **semimetals**. These elements are boron, silicon, germanium, arsenic, antimony, tellurium and astatine.

The most metallic element is francium at the bottom left of the table. The most nonmetallic element is fluorine. The metallic character of elements within a group increases with period number. This means that **within a column, the more metallic elements are at the bottom**. The metallic character of elements within a period decreases with the increase in group number. This means that **within a row, the more metallic elements are on the left**.

Among the main group atoms, **elements diagonal to each other** as indicated by the dashed arrows **have similar properties** because they have a similar metallic character. The noble gases are nonmetals, but they are an exception to the diagonal rule.

Physical properties relating to metallic character are summarized in the following table:

Element	Electrical/thermal conductivity	Malleable/ductile as solids?	Lustrous?	Melting point of oxides, hydrides, and halides
Metals	High	Yes	Yes	High
Metalloids	Intermediate. Altered by dopants (semiconductors)	No (brittle)	Varies	Varies (oxides). Low (hydrides, halides)
Nonmetals	Low (insulators)	No	No	Low

Malleable materials can **be beaten into sheets**. **Ductile** materials can **be pulled into wires**. **Lustrous** materials **have a shine**. Oxides, hydrides, and halides are compounds with O, H, and halogens respectively. Measures of intermolecular attractions other than melting point are also higher for metal oxides, hydrides, and halides than for the nonmetal compounds. A dopant is a small quantity of an intentionally added impurity. The controlled movement of electrons in doped silicon semiconductors carries digital information in computer circuitry.

To compare the size of **an atom,** we need to compare radii among different atoms using some standard. As seen to the right, the sizes of atoms decrease with period number and increase with group number. This trend is similar to the trend described above for metallic character. The smallest atom is helium.

size increase

Group names, melting point, density, and properties of compounds

Groups 1, 2, 17, and 18 are often identified with a **group name**. These names are shown in the table below. Several elements are found as **diatomic molecules**: **(H_2, N_2, O_2, and the halogens: F_2, Cl_2, Br_2, and I_2)**. Mnemonic devices to remember the diatomic elements are: "$Br_2I_2N_2Cl_2H_2O_2F_2$" (pronounced "Brinklehof" and "**H**ave **N**o **F**ear **O**f **I**ce **C**old **B**eer." These molecules are attracted to one another using **weak London dispersion forces**.

Note that **hydrogen** is not an alkali metal. Hydrogen is a colorless gas and is the most abundant element in the universe, but H_2 is very rare in the atmosphere because it is light enough to escape gravity and reach outer space. Hydrogen atoms form more compounds than any other element.

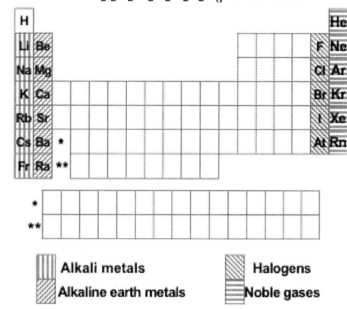

Alkali metals are shiny, soft, metallic solids. They have **low melting points and low densities** compared with other metals (see squares in figures on the following page) because they have a weaker metallic bond. Measures of intermolecular attractions including their **melting points decrease further down the periodic table due to weaker metallic bonds** as the size of atoms increases.

Alkaline earth metals (group 2 elements) are grey, metallic solids. They are harder, denser, and have a higher melting point than the alkali metals (see asterisks in figures on the following page), but values for these properties are still low compared to most of the transition metals. Measures of metallic bond strength like melting points for alkaline earths do not follow a simple trend down the periodic table.

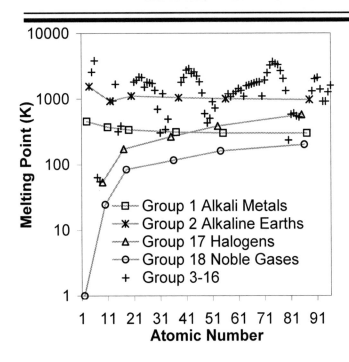

Halogens (group 17 elements) have an irritating odor. Unlike the metallic bonds between alkali metals, **London forces between halogen molecules increase in strength further down the periodic table**. Their melting points increase as shown by the triangles to the left. London forces make Br_2 a liquid and I_2 a solid at 25 °C. The lighter halogens are gases.

Noble gases (group 18 elements) have no color or odor and exist as **individual gas atoms** that experience London forces. These attractions also increase with period number as shown by the circles to the left.

The known **densities** of liquid and solid elements at room temperature are shown to the right. **Intermolecular forces contribute to density** by bringing nuclei closer to each other, so the periodicity is similar to trends for melting point. These group-to-group differences are superimposed on a general trend for **density** to **increase with period number** because heavier nuclei make the material denser.

Trends among properties of **compounds** may often be deduced from **trends among their atoms**, but caution must be used. For example, the densities of three potassium halides are:

2.0 g/cm^3 for KCl   2.7 g/cm^3 for KBr
3.1 g/cm^3 for KI.

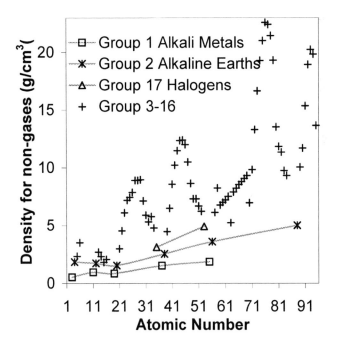

We would expect this trend for increasing atomic mass within a group. We might also expect the density of KF to be less than 2.0 g/cm^3, but it is actually 2.5 g/cm^3 due to a change in crystal lattice structure.

Physics of electrons and stability of electron configurations

For an isolated atom, the **most stable system of valence electrons is a filled set of orbitals**. For the main group elements, this corresponds to group 18 (ns^2np^6 and $1s^2$ for helium), and, to a lesser extent, group 2 (ns^2). The next most stable state is a set of degenerate half-filled orbitals. These occur in group 15 (ns^2np^3). The least stable valence electron configuration is a single electron with no other electrons in similar orbitals. This occurs in group 1 (ns^1) and to a lesser extent in group 13 (ns^2np^1).

An atom's first **ionization energy** is the energy required to remove one electron by the reaction

$M(g) \rightarrow M^+(g) + e^-$.

Periodicity is in the opposite direction from the trend for atomic radius. The most metallic atoms have electrons further from the nucleus, and these are easier to remove.

An atom's **electron affinity** is the energy released when one electron is added by the reaction

$M(g) + e^- \rightarrow M^-(g)$. A large negative number for the exothermic reaction indicates a high electron affinity. Halogens have the highest electron affinities.

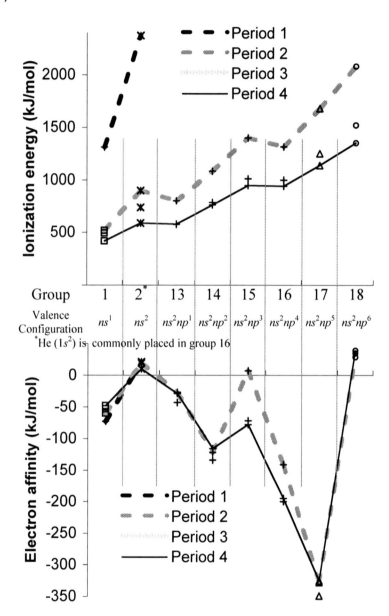

Group	1	2*	13	14	15	16	17	18
Valence Configuration	ns^1	ns^2	ns^2np^1	ns^2np^2	ns^2np^3	ns^2np^4	ns^2np^5	ns^2np^6

*He ($1s^2$) is commonly placed in group 16

Trends in **ionization energy and electron affinity** within a period reflect the **stability of valence electron configurations**. A stable system requires more energy to change and releases less when changed. Note the peaks in stability for groups 2, 13, and 16.

Much of chemistry consists of atoms **bonding** to achieve stable valence electron configurations. **Nonmetals gain electrons or share electrons** to achieve these configurations and **metals lose electrons** to achieve them.

Qualitative group trends

When cut by a knife, the exposed surface of an **alkali metal or alkaline earth metal** quickly turns into an oxide. These elements **do not occur in nature as free metals**. Instead, they react with many other elements to form white or grey water-soluble salts. With some exceptions, the oxides of group 1 elements have the formula M_2O, their hydrides are MH, and their halides are MX (for example, NaCl). The oxides of group 2 elements have the formula MO, their hydrides are MH_2, and their halides are MX_2, for example, $CaCl_2$.

Halogens form a wide variety of oxides and also combine with other halogens. They combine with hydrogen to form HX gases, and these compounds are also commonly used as acids (hydrofluoric, hydrochloric, etc.) in aqueous solution. Halogens form salts with metals by gaining electrons to become X^- ions. Astatine is an exception to many of these properties because it is an artificial metalloid.

Noble gases are **nearly chemically inert**. The heavier noble gases form a number of compounds with oxygen and fluorine such as KrF_2 and XeO_4

Electronegativity and reactivity series

Electronegativity measures the ability of an atom to attract electrons in a chemical bond. The most metallic elements have the lowest electronegativity. The most nonmetallic have the highest electronegativity.

In a reaction with a metal, the most reactive chemicals are the **most electronegative elements** or compounds containing those elements. In a reaction with a nonmetal, the most reactive chemicals are the **least electronegative elements** or compounds containing them. The reactivity of elements may be described by a **reactivity series**: an ordered list with chemicals that react strongly at one end and nonreactive chemicals at the other. The following reactivity series is for metals reacting with oxygen:

Metal	K	Na	Ca	Mg	Al	Zn	Fe	Pb	Cu	Hg	Ag	Au
Reaction with O_2	Burns violently		Burns rapidly					Oxidizes slowly			No reaction	

Copper, silver, and gold (group 11) are known as the **noble metals** or **coinage metals** because they rarely react.

<u>Valence and oxidation numbers</u>

The term **valence** is often used to describe the number of atoms that may react to form a compound with a given atom by sharing, removing, or losing **valence electrons**. A more useful term is **oxidation number**. The **oxidation number of an ion is its charge**. The oxidation number of an atom sharing its electrons is **the charge it would have if the bonding were ionic**. There are four rules for determining oxidation number:

1) The oxidation number of an element (i.e., a Cl atom in Cl_2) is zero because the electrons in the bond are shared equally.
2) In a compound, the more electronegative atoms are assigned negative oxidation numbers and the less electronegative atoms are assigned positive oxidation numbers equal to the number of shared electron-pair bonds. For example, hydrogen may only have an oxidation number of −1 when bonded to a less electronegative element or +1 when bonded to a more electronegative element. Oxygen almost always has an oxidation number of −2. Fluorine always has an oxidation number of −1 (except in F_2).
3) The oxidation numbers in a compound must add up to zero, and the sum of oxidation numbers in a polyatomic ion must equal the overall charge of the ion.
4) The charge on a polyatomic ion is equal to the sum of the oxidation numbers for the species present in the ion. For example, the sulfate ion, SO_4^{2-}, has a total charge of -2. This comes from adding the -2 oxidation number for 4 oxygen (total -8) and the +6 oxidation number for sulfur.

Example: What is the oxidation number of nitrogen in the nitrate ion, NO_3^-?
Oxygen has the oxidation number of −2 (rule 2), and the sum of the oxidation numbers must be −1 (rule 3). The oxidation number for N may be found by solving for x in the equation $x + 3 \times (-2) = -1$. The oxidation number of N in NO_3^- is +5.

There is a **periodicity in oxidation numbers** as shown in the table below for examples of oxides with the maximum oxidation number. Remember that an element may occur in different compounds in several different oxidation states.

Group	1	2	13	14	15	16	17	18
Oxide with maximum oxidation number	Li_2O Na_2O	BeO MgO	B_2O_3 Al_2O_3	CO_2 SiO_2	N_2O_5 P_2O_5	SO_3	Cl_2O_7 Br_2O_7	XeO_4
Oxidation number	+1	+2	+3	+4	+5	+6	+7	+8

They are called "oxidation numbers" because oxygen was the element of choice for reacting with materials when modern chemistry began, and the result was Mendeleev arranging his first table to look similar to this one.

Acidity/alkalinity of oxides

Metal oxides form basic solutions in water because the ionic bonds break apart and the O^{2-} ion reacts to form hydroxide ions:

metal oxide $\rightarrow$ metal cation$(aq) + O^{2-}(aq)$ and $O^{2-}(aq) + H_2O(l) \rightarrow 2\,OH^-(aq)$

Ionic oxides containing a large cation with a low charge (Rb_2O, for example) are most soluble and form the strongest bases.

Covalent oxides form acidic solutions in water by reacting with water. For example:

$$SO_3(l) + H_2O(l) \rightarrow H_2SO_4(aq) \rightarrow H^+(aq) + HSO_4^-(aq)$$
$$Cl_2O_7(l) + H_2O(l) \rightarrow 2HClO_4(aq) \rightarrow 2H^+(aq) + 2ClO_4^-(aq)$$

Covalent oxides at high oxidation states and high electronegativities form the strongest acids. For this skill, note that the periodic trends for acid and base strength of the oxide of an element follows the same pattern we've seen before.

An **electron configuration** is a **list of subshells** with superscripts representing the **number of electrons** in each subshell.

Electrons are found outside the nucleus in a fuzzy area called the electron cloud. We don't know exactly where the electron is but we do know the most probable place to find an electron with a certain energy. This is an orbital or an energy level.

When an electron is in its unexcited or ground state, there are seven energy levels. These seven energy levels match the seven periods (rows) of the Periodic Table. The valence, or outermost electrons are found in the energy level that corresponds to the period number.

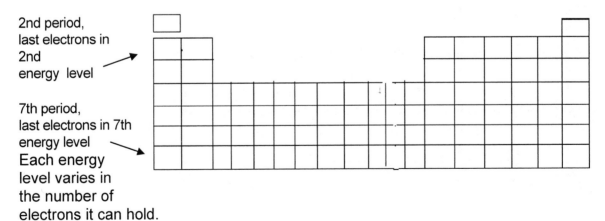

2nd period, last electrons in 2nd energy level

7th period, last electrons in 7th energy level
Each energy level varies in the number of electrons it can hold.

Within each energy level, the electrons are arranged into various sublevels called orbitals. The orbitals of the unexcited state atom include s, p, d, and f orbitals. The electrons fill these orbitals in a pre-determined pattern; filling the lowest energy orbitals first. "S" orbitals are the lowest energy so they fill with a maximum of two electrons first, followed by the "p" orbitals. There are three different "p" orbitals, each holding up to two electrons on each energy level. There are five different "d" orbitals followed by seven different "f" orbitals. Filling of the "d" and "f' orbitals is very complicated. The 3 d orbitals are

Energy level	Maximum Number of Electrons
1	2
2	8
3	18
4	32
5	50, theoretical, not filled
6	72, theoretical, not filled
7	98, theoretical, not filled

of higher energy than the 4s but less energy than the 4 p so they fill between the 4 s and 4 p orbitals, completing the third energy level. The same holds true for the s, p, and d orbitals on the 4th, 5th and 6th energy levels. The 4f orbital has more energy than the 6s but less energy than the 6p or 5d orbitals so it fills between the 6s and the 5d orbitals. This pattern really does make sense.

*However, there exceptions to this general rule, which is n squared times 2 (e.g., energy level,1 has 2 electrons, n=number of energy level, i.e., 1times 1 = 1, and 1 times 2 = 2). In K and Ca, there are 8 electrons in energy level 3.Ca – 2,8,8,2 and k – 2,8,8,1.

According to the Pauli Exclusion Principle, and Hund's Rule for filling electron orbitals, each orbital of the same type must fill with two electrons, spinning in opposite directions, before a new type of orbital is occupied. However, before electrons can double up in an orbital, all orbitals on the same energy level of the same type must have one electron in it, all spinning in the same direction. An electron of an atom can be identified using a shorthand notation called electron configuration. This gives the energy level, orbital type and number of electrons present.

For example, carbon has six electrons and they are located as follows:
    1st energy level, s orbital: 2 electrons,
    2nd energy level, 4 electrons; 2 in the s orbital and 2 in the p orbital.

For an electron configuration of: $1s^2, 2s^2, 2p^4$.

Chlorine has 17 electrons:
  1st energy level: 2 in the s orbital,
  2nd energy level: 2 in the s orbital, 6 in the p orbital
  3rd energy level: 2 in the s orbital, 5 in the p orbital.

For an electron configuration of: $1s^2, 2s^2, 2p^6, 3s^2, 3p^5$

Let's try a hard one: lead has 82 electrons.
  1st energy level: 2 in the s orbital,
  2nd energy level: 2 in the s orbital, 6 in the p orbital
  3rd energy level: 2 in the s orbital, 6 in the p orbital.

Then the 4s orbital fills with 2 electrons followed by the 3d orbital with 10 electrons then the 4p orbital with 6 electrons.
Now the 5s fills with 2 electrons, followed by the 4d orbital with 10 electrons, and then the 5p orbital with 6 electrons.
The sixth energy starts to fill next with 2 electrons in the 6s orbital, followed by 14 in the 4f and 10 electrons in the 5d and two electrons in the 6p orbital.
For an electron configuration of: $1s^2, 2s^2,$
$2p^6, 3s^2, 3p^6 4s^2 3d^{10} 4p^6 5s^2 4d^{10} 5p^6 6s^2 4f^{14} 5d^{10} 6p^2$

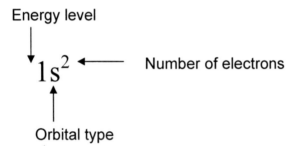

Energy level

$1s^2$ ⟵ Number of electrons

Orbital type

There is an electron configuration pattern that coincides with the Periodic Table. The last electrons in the electron cloud of all alkali and alkali earth metals are found in the "s" orbitals.

Element	Last electron to fill	Element	Last electron to fill
Li	$2s^1$	Be	$2s^2$
Na	$3s^1$	Mg	$3s^2$
K	$4s^1$	Ca	$4s^2$

Groups 13-18 have their last electrons in the "P" orbitals. Transition metals, groups 3-12, have their last electrons in the "d" orbitals. The lanthanide and actinide series have their last electrons in the "f" orbital.

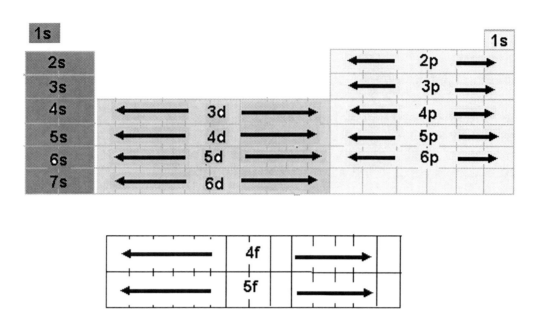

Nonmetals gain electrons or share electrons to achieve stable configurations and **metals lose electrons** to achieve them.

Valence electrons are the electron involved in chemical reactions so let's take a look at the relationship between the valence electron configuration and chemical activity. Remember the octet rule, it seems logical that when there are less than four valence electrons, the atom would want to lose electrons to go back to the previous energy level that was full. And that when the outer energy level is more than half full (more than 4 valence electrons) that the atom would want to gain electrons to complete the valence energy level.

For example: aluminum has 13 electrons with a configuration of $1s^2, 2s^2, 2p^6, 3s^2, 3p^1$.
There are only three electrons in the 3rd energy level and the octet rule says there should be eight. Aluminum can try and find 5 more electrons to fill the 3 p orbital or lose the $3s^2$, and $3p^1$ electrons. It takes less energy to lose the three electrons (circled), When these electrons are lost an aluminum +3 ion forms with an electron configuration of $1s^2, 2s^2, 2p^6$ which has a complete octet in the outer shell. Al lost 3 electrons and now Al has 10 electrons and 13 protons and the charge is 3+.

Keeping this in mind, along with the electron configuration chart above, it seems that the families with valence configurations s^1, s^2 or s^2, p^1 would tend to lose electrons to form ionic compounds. Families with valence configurations s^2, p^2; s^2, p^3; s^2, p^4; s^2, p^5 would tend to gain electrons to form ionic compounds or share electrons to form molecular substances. The noble gas family with a s^2, p^6 configuration has a complete octet in its valence shell so this configuration tends to be nonreactive. However, the heavier noble gases form a number of compounds with oxygen and fluorine such as KrF_2 and XeO_4.

Don't forget that atoms with electrons in the "d" or "f" orbitals, have s^1 or s^2 as their valence configuration of electrons.

This information helps us to understand some other characteristics.
For example, the elements in the alkali metal family, for example, are not found as free elements in nature. This is due to their high chemical reactivity. The one valence electron in the outermost energy level is high unstable, and it tends to be easily lost, forming many different ionic compounds in the process. The halogen family consists of molecules instead of free elements. The almost complete valence energy level is easily completed by two halogen atoms sharing electrons to form a covalent molecule.

In a reaction with a metal, the most reactive chemicals are the **most electronegative elements** or compounds containing those elements. In a reaction with a nonmetal, the most reactive chemicals are the **least electronegative elements** or compounds containing them. The reactivity of elements may be described by a **reactivity series**: an ordered list with chemicals that react strongly at one end and nonreactive chemicals at the other. The following reactivity series is for metals reacting with oxygen:

Metal	K	Na	Ca	Mg	Al	Zn	Fe	Pb	Cu	Hg	Ag	Au
Reaction with O_2	Burns violently		Burns rapidly					Oxidizes slowly			No reaction	

Summary

A summary of periodic trends is shown to the right. The properties tend to decrease or increase as shown depending on a given element's proximity to fluorine in the table.

http://jcrystal.com/ steffenweber/JAV A/jpt/jpt.html contains an applet of the periodic table and trends.

http://www.webele ments.com is an on-line reference for information on the elements.

http://www.uky.edu/Projects/Chemcomics/ has comic book pages for each element.

Metallic character ⇓   Ionization energy ⇑
Ionic character of halides ⇓   Covalent character of halides ⇑
Atomic radius ⇓   Electroneativity ⇑
Alkalinity of oxides ⇓   Acidity of oxides ⇑

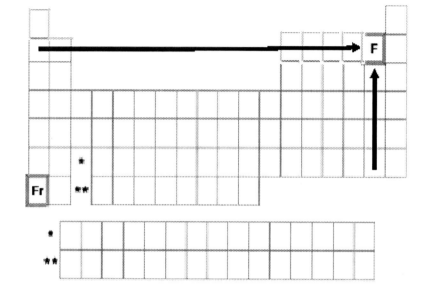

COMPETENCY 0008   UNDERSTAND THE KINETIC THEORY, THE NATURE OF PHASE CHANGES, AND THE GAS LAWS.

Molecules have **kinetic energy** (they move around), and they also have **intermolecular attractive forces** (they stick to each other). The relationship between these two determines whether a collection of molecules will be a gas, liquid, or solid.

A **gas** has an indefinite shape and an indefinite volume. The kinetic model for a gas is a collection of widely separated molecules, each moving in a random and free fashion, with negligible attractive or repulsive forces between them. Gases will expand to occupy a larger container so there is more space between the molecules. Gases can also be compressed to fit into a small container so the molecules are less separated.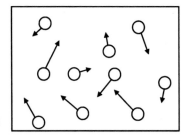
Diffusion occurs when one material spreads into or through another. Gases diffuse rapidly and move from one place to another.

A **liquid** assumes the shape of the portion of any container that it occupies and has a specific volume. The kinetic model for a liquid is a collection of molecules attracted to each other with sufficient strength to keep them close to each other but with insufficient strength to prevent them from moving around randomly. Liquids have a higher density and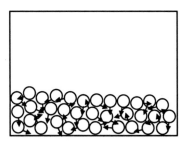
can not be compressed because the molecules in a liquid are closer together. Diffusion occurs more slowly in liquids than in gases because the molecules in a liquid stick to each other and are not completely free to move.

A **solid** has a definite volume and definite shape. The kinetic model for a solid is a collection of molecules attracted to each other with sufficient strength to essentially lock them in place. Each molecule may vibrate, but it has an average position relative to its neighbors. If these positions form an ordered pattern, the solid is called **crystalline**.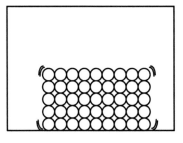
Otherwise, it is called **amorphous**. Solids have a high density and are almost incompressible because the molecules are close together. Diffusion occurs extremely slowly because the molecules almost never alter their position.

In a solid, the energy of intermolecular attractive forces is much stronger than the kinetic energy of the molecules, so kinetic energy and kinetic molecular theory are not very important. As temperature increases in a solid, the vibrations of individual molecules grow more intense and the molecules spread slightly further apart, decreasing the density of the solid.

In a liquid, the energy of intermolecular attractive forces is about as strong as the kinetic energy of the molecules and both play a role in the properties of liquids.

In a gas, the energy of intermolecular forces is much weaker than the kinetic energy of the molecules. Kinetic molecular theory is usually applied for gases and is best applied by imagining ourselves shrinking down to become a molecule and picturing what happens when we bump into other molecules and into container walls.

Phase changes occur when the relative importance of kinetic energy and intermolecular forces is altered sufficiently for a substance to change its state.

The transition from gas to liquid is called **condensation** and from liquid to gas is called **vaporization**. The transition from liquid to solid is called **freezing** and from solid to liquid is called **melting**. The transition from gas to solid is called **deposition** and from solid to gas is called **sublimation**.

Heat removed from a substance during condensation, freezing, or deposition permits new intermolecular bonds to form, and heat added to a substance during vaporization, melting, or sublimation breaks intermolecular bonds. During these phase transitions, this **latent heat** is removed or added with **no change in the temperature** of the substance because the heat is not being used to alter the speed of the molecules or the kinetic energy when they strike each other or the container walls. Latent heat alters intermolecular bonds.

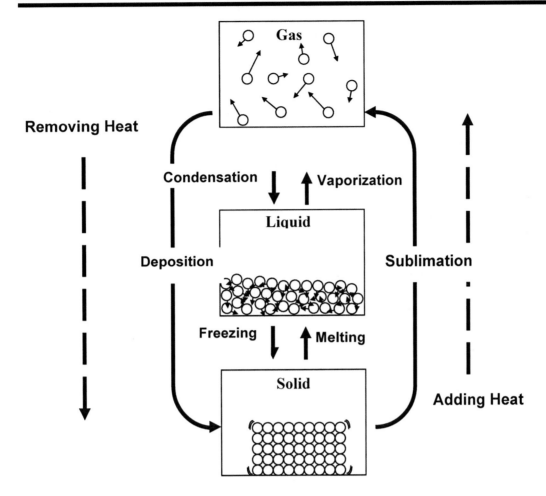

The kinetic molecular theory describes how an ideal gas behaves when conditions such as temperature, pressure, volume or quantity of gas are varied within a system. An **ideal gas** is an imaginary gas that obeys all of the assumptions of the kinetic molecular theory. The assumptions of the kinetic molecular theory help explain the behavior of gas molecules and their physical properties.

1. Gases are composed of a large number of particles that behave like hard, spherical objects in a state of constant, random motion.
2. These particles move in a straight line until they collide with another particle or the walls of the container.
3. These particles are much smaller than the distance between particles. Most of the volume of a gas is therefore empty space.
4. There is no force of attraction between gas particles or between the particles and the walls of the container.
5. Collisions between gas particles or collisions with the walls of the container are perfectly elastic. None of the energy of a gas particle is lost when it collides with another particle or with the walls of the container.
6. The average kinetic energy of a collection of gas particles depends on the temperature of the gas and nothing else.

While an ideal gas does not exist, most gases will behave like an ideal gas except when at very low temperatures or very high pressures. Under these conditions, gases vary from ideal behavior.

Charles's law states that the volume of a fixed amount of gas at constant pressure is directly proportional to absolute temperature, or:

$$V \propto T.$$

Or $V=kT$ where k is a constant. This gives a mathematical equation $\dfrac{V_1}{T_1} = \dfrac{V_2}{T_2}$.

Changes in temperature or volume can be found using Charles's law.

Example: What is the new volume of gas if 0.50 L of gas at 25ºC is allowed to heat up to 35ºC at constant pressure?

Solution: This is a volume-temperature change so use Charles's law. Temperature must be on the Kelvin scale. K= ºC + 273.

T_1= 298K
V_1= 0.50 L
T_2= 308K
V_2 =?

Use the equation: $\dfrac{V_1}{T_1} = \dfrac{V_2}{T_2}$ and rearrange for $V_2 = \dfrac{T_2 V_1}{T_1}$.

Substitute and solve V_2=0.52L.

Boyle's law states that the volume of a fixed amount of gas at constant temperature is inversely proportional to the gas pressure, or:

$$V \propto \frac{1}{P}.$$

Or $V=k/P$ where k is a constant. This gives a mathematical equation $P_1V_1=P_2V_2$. Pressure-volume changes can be determined using Boyle's law.

Example: A 1.5 L gas has a pressure of 0.56 atm. What will be the volume of the gas if the pressure doubles to 1.12 atm at constant temperature?

Solution: This is a pressure-volume relationship at constant temperature so use Boyle's law.

P_1= 0.56 atm
V_1= 1.5 L
P_2= 1.12 atm
V_2=?

Use the equation $P_1V_1=P_2V_2$, rearrange to solve for $V_2 = \dfrac{P_1V_1}{P_2}$.

Substitute and solve. V_2=0.75 L

Gay-Lussac's law states that the pressure of a fixed amount of gas in a fixed volume is proportional to absolute temperature, or:
$$P \propto T.$$

Or $P=kT$ where k is a constant. This gives a mathematical equation $\dfrac{P_1}{T_1} = \dfrac{P_2}{T_2}$.

Changes in temperature or pressure can be found using Gay-Lussac's law.

Example: A 2.25 L container of gas at 25°C and 1.0 atm pressure is cooled to 15°C. How does the pressure change if the volume of gas remains constant?
Solution: This is a pressure-volume change so use Gay-Lussac's law.

P_1= 1.0 atm
T_1= 25 °C
T_2= 15 °C

Change the temperatures to the Kelvin scale. K=°C+ 273.

Use the equation $\dfrac{P_1}{T_1} = \dfrac{P_2}{T_2}$ to solve. Rearrange the equation to solve for P_2, substitute and solve.

$P_2 = \dfrac{P_1 T_2}{T_1} = 0.97$ atm

The **combined gas law** uses the above laws to determine a proportionality expression that is used for a constant quantity of gas:

$$V \propto \frac{T}{P}.$$

The combined gas law is often expressed as an equality between identical amounts of an ideal gas at two different states ($n_1 = n_2$):

$$\frac{P_1 V_1}{T_1} = \frac{P_2 V_2}{T_2}.$$

Example: 1.5 L of a gas at STP is allowed to expand to 2.0 L at a pressure of 2.5 atm. What is the temperature of the expanded gas?

Since pressure, temperature and volume are changing use the combined gas law to determine the new temperature of the gas.

P_1= 1.0 atm
T_1= 273K
V_1= 1.5 L
V_2= 2.0L
P_2= 2.5 atm
T_2=?

Using this equation, $\dfrac{P_1 V_1}{T_1} = \dfrac{P_2 V_2}{T_2}$, rearrange to solve for T_2.

$$T_2 = \frac{P_2 V_2 T_1}{P_1 V_1}$$

Substitute and solve T_2=910 K or 637 °C

Avogadro's hypothesis states that equal volumes of different gases at the same temperature and pressure contain equal numbers of molecules. **Avogadro's law** states that the volume of a gas at constant temperature and pressure is directly proportional to the quantity of gas, or:

$$V \propto n, \text{ where } n \text{ is the number of moles of gas.}$$

Avogadro's law and the combined gas law yield $V \propto \dfrac{nT}{P}$.

The proportionality constant R--the **ideal gas constant**--is used to express this proportionality as the **ideal gas law**:

$$PV = nRT.$$

The ideal gas law ($PV = nRT$) is useful because it contains all the information of Charles's, Avogadro's, Boyle's, and the combined gas laws in a single expression.

If pressure is given in atmospheres and volume is given in liters, a value for R of **0.08206 L-atm/(mol-K)** is used. If pressure is given in Pascal (newtons/m^2) and volume in cubic meters, then the SI value for R of **8.314 J/(mol-K)** may be used because a joule is defined as a Newton-meter. A value for R of **8.314 m^3-Pa/(mol-K)** is identical to the ideal gas constant using joules.

Many problems are given at "**standard temperature and pressure**" or "**STP**." Standard conditions are *exactly* **1 atm** (101.325 kPa) and **0 °C (273.15 K)**. At STP, one mole of an ideal gas has a volume of:

$$V = \frac{nRT}{P}$$

$$= \frac{(1 \text{ mole})\left(0.08206 \ \frac{\text{L-atm}}{\text{mol-K}}\right)(273 \text{ K})}{1 \text{ atm}} = 22.4 \text{ L}.$$

The value of 22.4 L is known as the **standard molar volume of any gas at STP**.

Solving gas law problems using these formulas is a straightforward process of algebraic manipulation. **Errors commonly arise from using improper units**, particularly for the ideal gas constant R. An absolute temperature scale must be used (never °C) and is usually reported using the Kelvin scale, but volume and pressure units often vary from problem to problem. Temperature in Kelvin is found from:

$$T \text{ (in K)} = T(\text{in °C}) + 273.15$$

Tutorials for gas laws may be found online at: http://www.chemistrycoach.com/tutorials-6.htm. A flash animation tutorial for problems involving a piston may be found at http://www.mhhe.com/physsci/chemistry/essentialchemistry/flash/gasesv6.swf.

Example: What volume will 0.50 mole an ideal gas occupy at 20.0 °C and 1.5 atm?

Solution: Since the problem deals with moles of gas with temperature and pressure, use the ideal gas law to find volume.

$PV = nRT$ $V = nRT/P$

$P = 1.5$ atm $= 0.50$ mol $(0.0821$ atm L/mol K$)$ 293 K/

$V= ?$ 1.5 atm

$n= 0.50$ mol $= 8.0$ L

$T=20.0$ °C$= 293$ K

$R = 0.0821$ atm L/mol K

Example: At STP, 0.250 L of an unknown gas has a mass of 0.491 g. Is the gas SO_2, NO_2, C_3H_8, or Ar? Support your answer.

Solution: Identify what is given and what is asked to determine.

Given: $T_1= 273K$
        $P_1=1.0$ atm
        $V_1= 0.250$ L
        Mass$= 0.419$ g

Determine: Identity of the gas. In order to do this, must find molar mass of the gas. $n = \dfrac{mass}{MM}$. Find the number of moles of gas present using $PV=nRT$ and then determine the MM to compare to choices given in the problem.

Solve for n$=\dfrac{PV}{RT}$ $= 0.011$ moles

$MM= \dfrac{mass}{n} = 38.1$ g/mol

Compare to MM of SO_2 (96 g/mol), NO_2 (46 g/mol), C_3H_8 (44 g/mol and Ar (39.9 g/mol). It is closest to Ar, so the gas is probably Argon.

COMPETENCY 0009   APPLY THE CONVENTIONS OF CHEMICAL NOTATION AND REPRESENTATIONS.

Today, the 115 named elements are organized into what is known as the Periodic Table. The design of this table provides much information about a particular element or group of elements. The table consists of boxes arranged in columns and rows.

| 1 |
| H |
| Hydrogen |
| 1.0079 |

The numbers inside each box give additional information about the element. There are two different sets of numbers. The first set are consecutive whole numbers. These numbers are called the atomic numbers. Atomic numbers indicates the number of protons in an atom of that element. This is characteristic for each element.

The identity of an **element** depends on the **number of protons** in the nucleus of the atom. Atoms and ions of a given element that differ in number of neutrons and have a different mass are called **isotopes**. A nucleus with a specified number of protons and neutrons is called a **nuclide**, and a nuclear particle, either a proton or neutron, may be called a **nucleon**. The total number of nucleons is called the **mass number** and this number is a whole number and is calculated by rounding up the average atomic mass value, for hydrogen, the atomic mass is 1.008 amu (atomic mass units) and is rounded up and the mass number of hydrogen is 1. 1 amu is equivalent to 1/12th of the atomic mass of carbon.

The difference between the mass number and the relative average mass is that the mass number is the nearest whole number to the average atomic mass and represents the sum of the number of protons and number of neutrons.

Symbolically, this information is displayed in this manner:

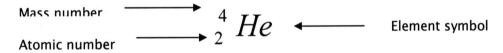

Mass number $\longrightarrow$ $^{4}_{2}He$ $\longleftarrow$ Element symbol

Atomic number $\longrightarrow$

This symbol provides a great deal of information about the composition of this particular atom. The atomic number, 2 in this case, tells the number of protons. The definition of an atom indicates that it is a neutral particle so the number of positively charged particles must be offset by the same number of negatively charged particles. Electrons are negatively charged, so the number of protons in an atom must be the same as the number of electrons in that atom. The mass number is defined as the number of protons plus the number of neutrons. Since the mass number of this example is 4, the sum of the number of protons and the number of neutrons is also 4. That means there are 2 neutrons in the nucleus. Hydrogen is the only element in with no neutrons since its nucleus has only one proton.

Different isotopes have different natural abundances and have different nuclear properties, but an atom's chemical properties are almost entirely due to electrons.

$$^{12}_{6}C$$ This symbol indicates the atomic number is 6 and that means that there are six protons. Since this is an atom, and all atoms are electrically neutral, there must also be six negatives charges or electrons to offset the +6 charge from the protons.

The symbol also indicates that the mass number is 12. The mass number is the nearest whole number to the actual atomic mass of the atom. Simply put, this is the number of protons plus the number of neutrons. Mass Number = $\#p^+ + \# n^0$. In this case, there are 6 p^+ and X n^0 that must add up to 12. So,

$$12 = 6 + x$$
$$x = 6$$

$$^{12}_{6}C \left\langle \begin{array}{l} \text{6 protons} \\ \text{6 electrons} \end{array} \right.$$

The mass number is the nearest whole number to the actual atomic mass of the atom. Simply put, this is the number of protons plus the number of neutrons. Mass Number = $\#p^+ + \# n^0$.

In this case, there are 6 p^+ and X n^0 that must add up to 12. So,

$$12 = 6 + x$$
$$x = 6$$

And there are 6 n^0

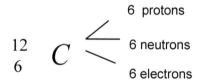

$$^{12}_{6}C \left\langle \begin{array}{l} \text{6 protons} \\ \text{6 neutrons} \\ \text{6 electrons} \end{array} \right.$$

In this isotope of carbon, things change a little: $$^{14}_{6}C$$

To summarize, there are 3 simple rules:
1. Atomic number = # of protons = # of electrons in an atom
2. Mass number = atomic mass rounded up to the nearest whole number
3. mass number = # of protons + # of neutrons

There are always 6 protons in carbon atoms, and because it is an atom, also 6 electrons.

But the mass number is now 14 meaning $6\,p^+ + X\,n^0 = 14$. The number of neutrons is 8 for this isotope of carbon.

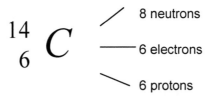

The IUPAC is the **International Union of Pure and Applied Chemistry**, an organization that formulates naming rules. **Organic compounds contain carbon**, and they have a separate system of nomenclature but some of the simplest molecules containing carbon also fall within the scope of inorganic chemistry.

Naming rules depend on whether the chemical is an ionic compound or a molecular compound containing only covalent bonds. There are special rules for naming acids. The rules below describe a group of traditional "semi-systematic" names accepted by IUPAC.

Ionic compounds:

Ionic compounds are named with the **cation (positive ion) first**. Nearly all cations in inorganic chemistry are **monatomic**, meaning they just consist of one atom (like Ca^{2+}, the calcium ion.) This atom will be a **metal ion**. For common ionic compounds, the **alkali metals always have a 1+ charge** and the **alkali earth metals always have a 2+ charge.**

Many metals may form cations of more than one charge. In this case, a Roman numeral in parenthesis after the name of the element is used to indicate the ion's charge in a particular compound.
Transition elements form ions with more than one charge, i.e. they have ions of different charges nd that is indicated as a Roman numeral in paranthesis next to the metal, e.g Fe (I) etc.

This Roman numeral method is known as the **Stock system**. An older nomenclature used the suffix –*ous* for the lower charge and –*ic* for the higher charge and is still used occasionally.

Example: Fe^{2+} is the iron(II) ion and Fe^{3+} is the iron(III) ion.

The only common inorganic **polyatomic cation** is **ammonium: NH_4^+.**

Ionic compounds:

The **anion** (negative ion) is named and written last. Monatomic anions are formed from nonmetallic elements and are named by **replacing the end of the element's name with the suffix –*ide.***

les: Cl⁻ is the chloride ion, S^{2-} is the sulfide ion, and N^{3-} is the nitride ion.

anions also end with –ide:

C_2^{2-}	N_3^-	O_2^{2-}	O_3^-	S_2^{2-}	CN⁻	OH⁻
carbide or acetylide	azide	peroxide	ozonide	disulfide	cyanide	hydroxide

Oxoanions (also called oxyanions) **contain one element in combination with oxygen**. Many common polyatomic anions are oxoanions that **end with the suffix –ate.** If an element has two possible oxoanions, the one with the element at a lower oxidation state **ends with –ite**. This anion will also usually have **less oxygen per atom**. Additional oxoanions are named with the prefix *hypo-* if they have a lower oxidation number than the *–ite* form and the prefix *per–* if they have a higher oxidation number than the *–ate* form.

Common examples:

				CO_3^{2-}	carbonate		
		SO_3^{2-}	sulfite	SO_4^{2-}	sulfate		
		PO_3^{3-}	phosphite	PO_4^{3-}	phosphate		
$N_2O_2^{2-}$	hyponitrite	NO_2^-	nitrite	NO_3^-	nitrate		
ClO⁻	hypochlorite	ClO_2^-	chlorite	ClO_3^-	chlorate	ClO_4^-	perchlorate
BrO⁻	hypobromite	BrO_2^-	bromite	BrO_3^-	bromate	BrO_4^-	perbromate
				MnO_4^{2-}	manganate	MnO_4^-	permanganate
				CrO_4^{2-}	chromate	CrO_8^{3-}	perchromate

Note that manganate/permanganate and chromate/perchromate are exceptions to the general rules because there are *–ate* ions but no *–ite* ions and because the charge changes.

Other polyatomic anions that end with *–ate* are:

$C_2O_4^{2-}$	$Cr_2O_7^{2-}$	SCN⁻	HCO_2^-	$CH_3CO_2^-$
oxalate	dichromate	thiocyanate	formate	acetate

HCO_2^- and $CH_3CO_2^-$ are condensed **structural formulas** because they show how the atoms are linked together. Their molecular formulas would be CHO_2^- and $C_2H_3O_2^-$

If an H atom is added to a polyatomic anion with a negative charge greater than one, the word *hydrogen* or the prefix *bi-* are used for the resulting anion. If two H atoms are added, *dihydrogen* is used.

Examples: bicarbonate or hydrogen carbonate ion: HCO_3^-
           dihydrogen phosphate ion: $H_2PO_4^-$

Ionic compounds: Hydrates

Water molecules often occupy positions within the lattice of an ionic crystal. These compounds are called **hydrates**, and the water molecules are known as **water of hydration**. The water of hydration is added after a centered dot in a formula. In a name, a number-prefix (listed below for molecular compounds) indicating the number of water molecules is followed by the root –*hydrate*.

Ionic compounds: Putting it all together

We now have the tools to name most common salts given a formula and to write a formula for them given a name. To determine a formula given a name, the number of anions and cations that are needed to achieve a neutral charge must be found.

Example: Determine the formula of cobalt(II) phosphite octahydrate.

Solution: For the cation, find the symbol for cobalt (Co) and recognize that it is present as Co^{2+} ions from the Roman numerals. For the anion, remember the phosphite ion is $PO_3{}^{3-}$. A neutral charge is achieved with 3 Co^{2+} ions for every 2 $PO_3{}^{3-}$ ions. Add eight H_2O for water of hydration for the answer:

$$Co_3\left(PO_3\right)_2 \bullet 8H_2O.$$

Molecular compounds

Molecular compounds (compounds making up molecules with a neutral charge) are usually composed entirely of nonmetals and are named by placing the **less electronegative atom first**. See **Skill 3.3** for the relationship between electronegativity and the periodic table. The suffix –*ide* is added to the second, more electronegative atom, and prefixes indicating numbers are added to one or both names if needed.

Prefix	mono-	di-	tri-	tetra-	penta-	hexa-	hepta-	octa-	nona-	deca-
Meaning	1	2	3	4	5	6	7	8	9	10

The final "o" or "a" may be left off these prefixes for oxides.

The electronegativity requirement is the reason the compound with two oxygen atoms and one nitrogen atom is called nitrogen dioxide, NO_2 and <u>not</u> dioxygen nitride O_2N. The hydride of sodium is NaH, sodium hydride, but the hydride of bromine is HBr, hydrogen bromide (or hydrobromic acid if it's in aqueous solution). Oxygen is only named first in compounds with fluorine such as oxygen difluoride, OF_2, and fluorine is never placed first because it is the most electronegative element.

Examples: N_2O_4, dinitrogen tetroxide (or tetraoxide)
Cl_2O_7, dichlorine heptoxide (or heptaoxide)
ClF_5 chlorine pentafluoride

Acids

There are special naming rules for acids that correspond with the **suffix of their corresponding anion** if hydrogen were removed from the acid. Anions ending with –ide correspond to acids with the prefix *hydro*– and the suffix –*ic*. Anions ending with –ate correspond to acids with no prefix that end with –*ic*. Oxoanions ending with –ite have associated acids with no prefix and the suffix –*ous*. The *hypo*– and *per*– prefixes are maintained. Some examples are shown in the following table:

anion	anion name	acid	acid name
Cl^-	chloride	$HCl(aq)$	hydrochloric acid
CN^-	cyanide	$HCN(aq)$	hydrocyanic acid
CO_3^{2-}	carbonate	$H_2CO_3(aq)$	carbonic acid
SO_3^{2-}	sulfite	$H_2SO_3(aq)$	sulfurous acid
SO_4^{2-}	sulfate	$H_2SO_4(aq)$	sulfuric acid
ClO^-	hypochlorite	$HClO(aq)$	hypochlorous acid
ClO_2^-	chlorite	$HClO_2(aq)$	chlorous acid
ClO_3^-	chlorate	$HClO_3(aq)$	chloric acid
ClO_4^-	perchlorate	$HClO_4(aq)$	perchloric acid

Example: What is the molecular formula of phosphorous acid?

Solution: If we remember that the –*ous* acid corresponds to the –*ite* anion, and that the –*ite* anion has one less oxygen than (or has an oxidation number 2 less than) the –*ate* form, we only need to remember that phosphate is PO_4^{3-}. Then we know that phosphite is PO_3^{3-} and phosphorous acid is H_3PO_3.

Properly written and named formulas

Proper formulas will follow the rules of the previous skill. Here are some ways to identify improper formulas that are emphasized below by underlining them.

In all common names for **ionic compounds, number prefixes are not used** to describe the number of anions and cations.

Examples: $CaBr_2$ is calcium bromide, <u>not calcium dibromide.</u>
$Ba(OH)_2$ is barium hydroxide, <u>not barium dihydroxide.</u>
Cu_2SO_4 is copper(I) sulfate, <u>not dicopper sulfate or copper(II) sulfate</u>
<u>or</u>
<u>dicopper sulfur tetroxide.</u>

All ionic compounds must have a **neutral charge in their formula** representations.

Example: <u>MgBr is an improperly written formula</u> because Mg ion always exists as 2+ and Br ion is always a 1– ion. $MgBr_2$, magnesium bromide, is correct.

Proper oxoanions and acids use the correct prefixes and suffixes.

Example: HNO_3 is nitric acid because NO_3^- is the nitrate ion.

In both ionic and molecular compounds, the **less electronegative element comes first**.

Example: <u>CSi is an improperly written formula</u> because Si is below C on the periodic table and therefore less electronegative. SiC, silicon carbide, is correct.

For additional resources, see:

<u>http://chemistry.alanearhart.org/Tutorials/Nomen/nomen-part7.html</u> has <u>thousands</u> of sample questions. Don't do them all in one sitting.

<u>http://www.iupac.org/reports/provisional/abstract04/connelly_310804.html</u> - IUPAC's latest report on inorganic nomenclature

Lewis dot structures/diagrams are a method for keeping track of each atom's valence electrons in a molecule. Drawing Lewis structures is a three-step process:

1) Add the number of valence shell electrons for each atom. If the compound is an anion, add the charge of the ion to the total electron count because anions have "extra" electrons. If the compound is a cation, subtract the charge of the ion (an easy way to remember is that the number of valence electrons for groups 1 7 2 is the group number, e.g., , H-1, Ca-2, etc. and for groups 13-18 it is group number minus 1, e.g Al-13, valence electron number is 3).

2) Write the symbols for each atom showing how the atoms connect to each other.

3) Draw a single bond (one pair of electron dots or a line) between each pair of connected atoms. Place the remaining electrons around the atoms as unshared pairs. If every atom has an octet of electrons except H, He, Li, and Be, which are atoms with two electrons, the Lewis structure is complete. Shared electrons count towards both atoms. If there are too few electron pairs to do this, draw multiple bonds (two or three pairs of electron dots between the atoms) until an octet is around each atom (except H atoms with two). If there are two many electron pairs to complete the octets with single bonds then the octet rule Is broken for this compound.

Example: Draw the Lewis structure of HCN.
Solution:
1) From their locations in the main group of the periodic table, we know that each atom contributes the following number of electrons: H—1, C—4, N—5. Because it is a neutral compound, the molecule will have a total of 10 valence electrons.

2) The atoms are connected with C at the center and will be drawn as:
H C N.

Having H as the central atom is impossible because H has one valence electron and will always only have a single bond to one other atom. If N were the central atom then the formula would probably be written as HNC.

$$\text{H} : \overset{\cdot\cdot}{\text{C}} : \overset{\cdot\cdot}{\underset{\cdot\cdot}{\text{N}}}$$

3) Connecting the atoms with 10 electrons in single bonds gives the structure to the right. H has two electrons to fill its valence subshells, but C and N only have six each. A triple bond between these atoms fulfills the octet rule for C and N and is the correct Lewis structure.

$$\text{H} : \text{C} ::: \text{N} :$$

To select the most probable Lewis dot structure for a compound or molecule that follows the octet rule, review the structures and compare to the method for constructing Lewis dot structures from the previous page.

Example: Which of the electron-dot structures given below for nitrous oxide (laughing gas), N_2O, is/are acceptable?

I. :N::N:O:

II. :N: N::O:

III. :N:::N::O:

Solution: Both nitrogen and oxygen follow the octect rule so the Lewis structure should show each atom in the molecule with 8 electrons, either unshared or shared. Upon examination, only choice I provides each atom in the molecule with 8 electrons. Choice II has only 6 electrons around each of the nitrogen atoms and choice III has 10 electrons around the center nitrogen atom.

Molecular geometry is predicted using the valence-shell/outermost energy level electron-pair repulsion or **VSEPR** model. VSEPR uses the fact that **electron pairs around the central atom of a molecule repel each other**. Imagine you are one of two pairs of electrons in bonds around a central atom (like a bonds in BeH_2 in the table below). You want to be as far away from the other electron pair as possible, so you will be on one side of the atom and the other pair will be on the other side. There is a straight line (or a 180° angle) between you to the other electron pair on the other side of the nucleus. In general, electron pairs lie at the **largest possible angles** from each other.

Electron pairs	Geometrical arrangement		Predicted bond angles	Example
2		Linear	180°	
3		Trigonal planar	120°	
4		Tetrahedral	109.5°	
5		Trigonal bipyramidal	120° and 90°	
6		Octahedral	90°	

X represents a generic central atom. Lone pair electrons on F are not shown in the example molecules.

Unshared Electron Pairs

The **shape of a molecule is given by the location of its atoms**. These a
connected to central atoms by shared electrons, but unshared electrons also
have an important impact on molecular shape. Unshared electrons may
determine the angles between atoms. Molecular shapes in the following table
take into account total and unshared electron pairs.

Electron pairs	Molecular shape				
	All shared pairs	1 unshared pair	2 unshared pairs	3 unshared pairs	4 unshared pairs
2	A—X—A Linear				
3	Trigonal planar	Bent			
4	Tetrahedral	Trigonal pyramidal	Bent		
5	Trigonal bipyramidal	Seesaw or sawhorse	T-shaped	Linear	
6	Octahedral	Square pyramidal	Square planar	T-shaped	Linear

X represents a generic central atom bonded to atoms labeled A.

Altered Bond Angles

Unpaired electrons also have a less dramatic impact on molecular shape.
Imagine you are an unshared electron pair around a molecule's central atom.
The shared electron pairs are each attracted partially to the central atom and
partially to the other atom in the bond, but you are different.

CHEMISTRY   95

You are attracted to the central atom, but there's nothing on your other side, so you are free to expand in that direction. That expansion means that you take up more room than the other electron pairs, and they are all squeezed a little closer together because of you. Multiple bonds have a similar effect because more space is required for more electrons. In general, **unshared electron pairs and multiple bonds decrease the angles between the remaining bonds**. A few examples are shown in the following tables.

Compound	CH_4	NH_3	H_2O
Unshared electrons	0	1	2
Shape	Tetrahedral	Trigonal pyramidal	Bent

Compound	BF_3	C_2H_4 (ethylene)
Multiple bonds	0	1
Shape	Trigonal planar	Trigonal planar

Summary

In order to use VSEPR to predict molecular geometry, perform the following steps:

1) Write out Lewis dot structures.
2) Use the Lewis structure to determine the number of unshared electron pairs and bonds around each central atom counting multiple bonds as one (for now).
3) The second table of this skill gives the arrangement of total and unshared electron pairs to account for electron repulsions around each central atom.
4) For multiple bonds or unshared electron pairs, decrease the angles slightly between the remaining bonds around the central atom.
5) Combine the results from the previous two steps to determine the shape of the entire molecule.

http://www.shef.ac.uk/chemistry/vsepr/ is a good site for explaining and visualizing molecular geometries using VSEPR.
http://cowtownproductions.com/cowtown/genchem/09_16T.htm provides some practice for determining molecular shape.

COMPETENCY 0010 UNDERSTAND THE PROCESS OF NUCLEAR TRANSFORMATION.

Some nuclei are unstable and emit particles and electromagnetic radiation. These emissions from the nucleus are known as **radioactivity**; the unstable isotopes are known as **radioisotopes**; and the nuclear reactions that spontaneously alter them are known as **radioactive decay**. Particles commonly involved in nuclear reactions are listed in the following table:

Particle	Neutron	Proton	Electron	Positron	Alpha particle	Beta particle	Gamma rays
Symbol	$_0^1n$	$_1^1p$ or $_1^1H$	$_{-1}^0e$	$_1^0e$	$_2^4\alpha$ or $_2^4He$	$_{-1}^0\beta$ or $_{-1}^0e$	$_0^0\gamma$

The main factor in determining whether a nucleus will decay is the neutron to proton ratio. Stable atoms have a neutron to proton ratio close to 1. The ratio in unstable atoms is greater than 1. Nature likes a balance between the neutrons and the protons, so the nuclei give off radioactivity in the form of alpha or beta particles in an attempt to bring the neutron to proton ratio closer to 1.

Ernest Rutherford classified this high energy radiation into three major types: Alpha particles, Beta particles and Gamma rays.

$_2^4He$

- Alpha particles are positively charged with a mass of 4 amu (atomic mass units, 1amu=1/12th of carbon atom). They are, then, the nuclei of Helium atoms. These particles are the slowest moving and the most massive of the three types of radioactivity. As such, they have the least penetrating ability through matter. A sheet of paper will stop their movement.

$_{-1}^0e$

- Beta particles were described by Rutherford as fast moving electrons. They are negatively charged with no protons or neutrons (or they have a mass of 0.) A beta particle is released when a neutron breaks into a proton and an electron. It is fast moving, high energy radiation and can penetrate paper and skin. A 3 mm thick sheet of metal is needed to stop it.

- Gamma radiation is electromagnetic radiation like light but with a greater frequency (10^7 hz for light compared to 10^{20} hz for gamma). It has no charge or mass.

$$^0_0\gamma$$

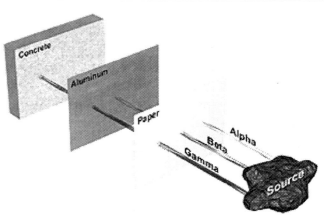

This radiation release often accompanies the release of alpha or beta particles because of the high energy arrangement left in the nucleus after the release of alpha or beta particles. Gamma radiation has the greatest penetrating ability. Concrete, 6 inches thick, is needed to stop gamma radiation.

Thorium-234 has 90 protons and 144 neutrons for a neutron to proton ratio greater than 1 (144 /90). Thorium-234 emits a beta particle. This particle is given off when a neutron breaks into a proton and an electron. The proton stays in the nucleus changing it to 91 protons which is Protactinium –234 while emitting the electron as a beta particle.

a neutron 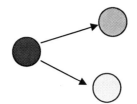 a proton which remains in the nucleus changing the atomic number of the substance, however

an electron seen as beta particle when emitted from the nucleus

So the change for Thorium-234 is:

$$^{234}_{90}\text{Th} \rightarrow {}^{0}_{-1}\text{e} + {}^{234}_{91}\text{Pa}$$

When an alpha particle is released by an unstable nucleus, the atom loses both mass and charge equal to that of a Helium nucleus; 4 amu and a +2 charge, ^4_2He.

Total mass the same on both sides of the equation: 238 = 4 + 234

$$^{238}_{92}\text{U} \rightarrow {}^{4}_{2}\text{He} + {}^{234}_{90}\text{Th}$$

Total charge the same on both sides of the equation: 92 = 2 + 90.

Nuclear transmutations are represented by nuclear equations. Nuclear equations show the change in the nucleus as well as the particle emitted during the decay process. Just like chemical equations, these equations must follow the Law of Conservation of Mass and the Law of Conservation of Charge. That is, they are balanced by equating the sum of mass numbers on both sides of a reaction equation and the sum of atomic numbers on both sides of a reaction equation.

The electron is assigned an atomic number of –1 to account for the conversion during radioactive decay of a neutron to a proton and an emitted electron called a **beta particle**:

$$_0^1n \rightarrow {_1^1}p + {_{-1}^0}e.$$

Sulfur-35 is an isotope that decays by beta emission:

$$_{16}^{35}S \rightarrow {_{17}^{35}}Cl + {_{-1}^0}e.$$

In most cases nuclear reactions result in a **nuclear transmutation** from one element to another. Transmutation was originally connected to the mythical "philosopher's stone" of alchemy that could turn cheaper elements into gold. When Frederick Soddy and Ernest Rutherford first recognized that radioactive decay was changing one element into another, Soddy remembered saying, "Rutherford, this is transmutation!" Rutherford replied, "Soddy, don't call it transmutation. They'll have our heads off as alchemists."

Isotopes may also decay by **electron capture** from an orbital outside the nucleus:

$$_{79}^{196}Au + {_{-1}^0}e \rightarrow {_{78}^{196}}Pt.$$

A **positron** is a particle with the small mass of an electron but with a positive charge. A positron emission converts a proton into a neutron. Carbon-11 decays by positron emission:

$$_6^{11}C \rightarrow {_5^{11}}B + {_1^0}e.$$

Large isotopes often decay by **alpha particle** emission:

$$_{92}^{238}U \rightarrow {_{90}^{234}}Th + {_2^4}He.$$

Gamma rays are high-energy electromagnetic radiation, and gamma radiation is almost always emitted when other radioactive decay occurs. Gamma rays usually aren't written into equations because neither the mass number nor the atomic number is altered. One exception is the annihilation of an electron by a positron, an event that only produces gamma radiation:

$$_{-1}^0e + {_1^0}e \rightarrow 2{_0^0}\gamma.$$

Example: Balance the following nuclear transmutation:

β decay

$$^{14}_{6}C \rightarrow ^{14}_{7}N + ^{0}_{-1}e$$

Solution: The sum of the mass numbers on both the left and right side of the arrow must be the same:

Left side	Right side
14	14

They are the same so the particle emitted during decay has a mass of 0.

The sum of the charge must be the same on the left side and right side of the arrow.

Left side	Right side
6	7

The right side has one too many positive charges to balance 6 positive charges on the left side. Adding -1 to the right side will make it balance with 6 positive charges.

So the charge of the particle emitted during decay is -1.

That particle is an electron. $^{0}_{-1}e$ and should be placed in the equation to complete.

$$^{14}_{6}C \rightarrow ^{14}_{7}N + ^{0}_{-1}e$$

Example: Complete the following nuclear transmutation:

$$^{234}_{90}Th \rightarrow ^{0}_{-1}e + \underline{^{234}_{91}Pa}$$

Solution: Again, the sum of the mass numbers on each side of the arrow must be the same as well as the sum of the charges on each side.

Left side	Right side
234	0

234 is needed on the right side to equal the left side.
and

Left side	Right side
90	-1

91 is needed on the right side to equal the left side so the particle that forms from the decay of this isotope is $^{234}_{91}Pa$ and should be inserted to complete the transmutation.

$$^{234}_{90}Th \rightarrow ^{0}_{-1}e + ^{234}_{91}Pa$$

Sometimes the nucleus can be changed by bombarding it with another type of particle. This is referred to as induced radioactivity. In 1934, Irene Curie, the daughter of Pierre and Marie Curie, and her husband, Frederic Joliot, announced the first synthesis of an artificial radioactive isotope. They bombarded a thin piece of aluminum foil with α-particles produced by the decay of polonium and found that the aluminum target became radioactive. Chemical analysis showed that the product of this reaction was an isotope of phosphorus.

$$^{27}_{13}Al + ^{4}_{2}He \rightarrow ^{30}_{15}P + ^{1}_{0}n$$

In the next 50 years, more than 2000 other artificial radionuclides were synthesized.

A shorthand notation has been developed for nuclear reactions such as the reaction discovered by Curie and Joliot. The parent (or target) nuclide and the daughter nuclide are separated by parentheses that contain the symbols for the particle that hits the target and the particle or particles released in this reaction.

$$^{27}_{13}Al(\alpha,n)^{30}_{15}P$$

Larger bombarding particles were eventually used to produce even heavier transuranium elements.

$$^{253}_{99}Es + ^{4}_{2}He \rightarrow ^{256}_{101}Md + ^{1}_{0}n$$

$$^{246}_{96}Cm + ^{12}_{6}C \rightarrow ^{254}_{102}No + 4\,^{1}_{0}n$$

The half-lives for α-decay and spontaneous fission decrease as the atomic number of the element increases. Element 104, for example, has a half-life for spontaneous fission of 0.3 seconds. Elements therefore become harder to characterize as the atomic number increases.

The **half-life** of a reaction is the **time required to consume half the reactant**. The rate of radioactive decay for an isotope is usually expressed as a half-life. Solving these problems is straightforward if the given amount of time is an exact multiple of the half-life. For example, the half-life of ^{233}Pa is 27.0 days. This means that 200 grams of ^{233}Pa will decay according to the following table:

Day	Number of half-lives	^{233}Pa remaining	^{233}Pa decayed since day 0
0	0	200 g	0 g
27.0	1	100 g	100 g
54.0	2	50.0 g	150.0 g
81.0	3	25.0 g	175.0 g
108.0	4	12.5 g	187.5 g

Regardless of whether the given amount of time is an exact multiple of the half-life, the following equation may be used:

$$A_{remaining} = A_{initially} \left(\frac{1}{2}\right)^{\frac{t}{t_{halflife}}}$$

where: $A_{remaining} \Rightarrow$ amount remaining

$A_{initially} \Rightarrow$ amount initially

$t \Rightarrow$ time

$t_{halflife} \Rightarrow$ half-life

Example: If we start with 1.000 g of Strontium-90, how much will remain after 5.00 years if the half-life of Sr-90 is 28.8 years?

Solution: Using the equation $A_{remaining} = A_{initially} (1/2)^{t/t\ halflife}$,

$A_{int} = 1.000$ g $(1/2)^{5.00/28.8} = 0.8866$ g

$$= 1.000 g \left(\frac{1}{2}\right)^{5.00/28.8}$$

$$= 0.8866 \text{ g remaining}$$

Example: An isotope of cesium (cesium-137) has a half-life of 30 years. If 1.0 mg of cesium-137 disintegrates over a period of 90 years, how many mg of cesium-137 would remain?

Solution: Using the equation

$$A_{remaining} = A_{initially} \left(\frac{1}{2}\right)^{\frac{t}{t_{halflife}}}$$

where: $A_{remaining} \Rightarrow$ amount remaining

$A_{initially} \Rightarrow$ amount initially

$t \Rightarrow$ time

$t_{halflife} \Rightarrow$ half-life

[handwritten:]
$t_{halflife} = 30$
$A_{initial} = 1.0$ mg
$t = 90$ yrs
$A = 1\,mg\left(\frac{1}{2}\right)^{90/30}$
$A = 0.125\,mg$

$A_{remaining} = X$
$A_{initially} = 1.0$ mg
t= 90 years
$t_{1/2} = 30$ years
then $A_{remaining} = 1.0$ mg $(1/2)^{90/30} = 0.125$ mg

Problem: A 2.5 gram sample of an isotope of strontium-90 was formed in a 1960 explosion of an atomic bomb at Johnson Island in the Pacific Test Site. The half-life of strontium-90 is 28 years. In what year will only 0.625 grams of this strontium-90 remain?

Solution: This is a little more challenging in that substituting into the above equation gives you: 0.625 g = 2.5 $(1/2)^{x/28}$. The solution requires simplifying and then taking the natural log of both sides:

ln (0.625/2.5) =ln 0.25= ln $(1/2^{x/28})$ x=56 years

Add 56 years to 1960 and the year will be 2016 when only 0.625 grams of Sr-90 remain from that test.

When two nuclei collide, they sometimes stick to each other and synthesize a new nucleus. This **nuclear fusion** was first demonstrated by the synthesis of oxygen from nitrogen and alpha particles:

$$^{14}_{7}N + ^{4}_{2}He \rightarrow ^{17}_{8}O + ^{1}_{1}H.$$

Fusion is also used to create new heavy elements, causing periodic tables to grow out of date every few years. In 2004, IUPAC approved the name roentgenium (in honor of Wilhelm Roentgen, the discoverer of X-rays) for the element first synthesized in 1994 by the following reaction:

$$^{209}_{83}Bi + ^{64}_{28}Ni \rightarrow ^{272}_{111}Rg + ^{1}_{0}n.$$

A heavy nucleus may also split apart into smaller nuclei by **nuclear fission.**

Nuclear power currently provides 17% of the world's electricity. Heat is generated by **nuclear fission of uranium-235 or plutonium-239**. This heat is then converted to electricity by boiling water and forcing the steam through a turbine. Fission of ^{235}U and ^{239}Pu occurs when **a neutron strikes the nucleus and breaks it apart into smaller nuclei and additional neutrons**. One possible fission reaction is:

$$^{1}_{0}n + ^{235}_{92}U \rightarrow ^{141}_{56}Ba + ^{92}_{36}Kr + 3\,^{1}_{0}n$$

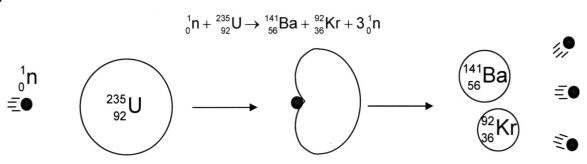

Gamma radiation, kinetic energy from the neutrons themselves, and the decay of the fission products (^{141}Ba and ^{92}Kr in the example above) all produce heat. The neutrons produced by the reaction strike other uranium atoms and produce more neutrons and more energy in a **chain reaction**. If enough neutrons are lost, the chain reaction stops and the process is called **subcritical**. If the mass of uranium is large enough so that one neutron on average from each fission event triggers another fission, the reaction is said to be **critical**. If the mass is larger than this so that few neutrons escape, the reaction is called **supercritical**. The chain reaction then multiplies the number of fissions that occur and the violent explosion of an atomic bomb will take place if the process is not stopped. The concentration of **fissile material** in nuclear power plants is sufficient for a critical reaction to occur but too low for a supercritical reaction to take place.

The alpha decay of **Plutonium-238 is used as a heat source for localized power generation** in space probes and in heart pacemakers from the 1970s.

The most promising nuclear reaction for producing power by nuclear fusion is:

$$^{2}_{1}H + ^{3}_{1}H \rightarrow ^{4}_{2}He + ^{1}_{0}n$$

Hydrogen-2 is called **deuterium** and is often represented by the symbol D. Hydrogen-3 is known as **tritium** and is often represented by the symbol T. Nuclear reactions between very light atoms similar to the reaction above are the energy source behind the sun and the hydrogen bomb.

Interconversion of mass and energy

With nuclear reactions, the energies involved are so great that the changes in mass become easily measurable. One no longer can assume that mass and energy are conserved separately, but must take into account their interconversion via Einstein's relationship, **E = mc²**.

If mass is in grams and the velocity of light is expressed as $c = 3 \times 10^{10}$ cm sec^{-1}, then the energy is in units of g cm^2 sec^{-2}, or ergs. A useful conversion is from mass in amu to energy in million electron volts (MeV):

1 amu = 931.4 MeV

What holds a nucleus together? If we attempt to bring two protons and two neutrons together to form a helium nucleus, we might reasonably expect the positively charged protons to repel one another violently. Then what keeps them together in the ${}^{4}_{2}\text{He}$ nucleus?

The answer is that a helium atom is lighter than the sum of two protons, two neutrons, and two electrons. Some of the mass of the separated particles is converted into energy and dissipates when the nucleus is formed. Before the helium nucleus can be torn apart into its component particles, this dissipated energy must be restored and turned back into mass. Unless this energy is provided, the nucleus cannot be taken apart. This energy is termed the *binding energy* of the helium nucleus.

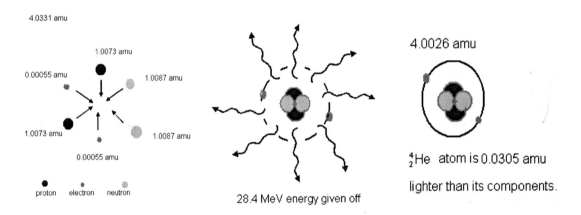

4.0331 amu

1.0073 amu

0.00055 amu   1.0087 amu

1.0073 amu   1.0087 amu

0.00055 amu

● proton   ※ electron   ※ neutron

28.4 MeV energy given off

4.0026 amu

${}^{4}_{2}\text{He}$ atom is 0.0305 amu

lighter than its components.

(We must include electrons in this calculation because 4.0026 amu is the mass of the helium-4 atom, not the nucleus.) This missing mass corresponds to 0.0305 x 931.4 MeV = 28.4 MeV of energy. If we could put together a helium atom directly from two neutrons, two protons, and two electrons, then 28.4 MeV of energy would be given off for every atom formed:

$$2p + 2n + 2e^- \longrightarrow {}^{4}_{2}\text{He} + 28.4 \text{ MeV of energy}$$

Compared to common chemical reactions, this is an enormous quantity of energy. Since 1 electron volt per atom is equivalent to 23.06 kcal per mole,

$$\text{binding energy} = 28.4 \text{ MeV atom}^{-1} \times \frac{23.06 \text{ kcal mole}^{-1}}{1 \text{ eV atom}^{-1}}$$

$$= 655,000,000 \text{ kcal mole}^{-1}$$

(Compare this energy with the 83 kcal mole^{-1} required to break carbon-carbon bonds in chemical reactions.)

Every nucleus is lighter than the sum of the masses of the nucleons from which it is built, and this loss of mass corresponds to the binding energy of the nucleus. The relative stability of two nuclei with different numbers of nucleons can be assessed by comparing their *loss of mass per nucleon.*

The loss of mass or binding energy per nuclear particle (protons and neutrons) rises rapidly to a maximum at iron, then falls. Iron is the most stable nucleus of all. The losses of masses or binding energies per nucleon are plotted below for all nuclei from helium through uranium. After some initial minor irregularities in the first- and second-row elements, the values settle down to a smooth curve, which rises to a maximum at iron, then begins a long descending slope through uranium and beyond.

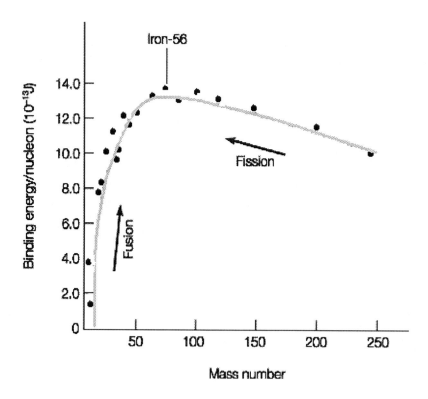

Iron is the most stable nucleus of all. For elements with smaller atomic numbers than iron, fusion of nuclei to produce heavier elements releases energy, because the products are lighter and more stable on a per-nucleon basis than the reactants. In contrast, beyond iron, fusion absorbs energy because the products are heavier on a per-nucleon basis than the reactants.

Example.
What is the loss of mass per nucleon for the $^{56}_{26}Fe$ atom, compared with its component protons, neutrons, and electrons?

Solution.
The $^{56}_{26}Fe$ atom contains 26 protons, 26 electrons, and 30 neutrons, so the mass calculation is performed as below:

Mass of 26 protons:	26 (1.0073) amu =	26.1898 amu
Mass of 30 neutrons:	30(1.0087) amu =	30.261 amu
Mass of 26 electrons:	26(0.00055) amu =	0.0143 amu
Total mass of components in iron-56		56.465 amu
Mass of iron-56 atom		55.93 amu
Total mass loss:		0.54 amu
Mass loss/nucleon:	0.54amu/56 =	0.0096 amu

Notice that the loss of mass per nucleon, and hence the binding energy per nucleon, is greater for iron (0.0096 amu) than for helium (0.0076 amu). This means that the iron nucleus is more stable relative to protons and neutrons than the helium nucleus is. If some combination of helium nuclei could be induced to produce an iron nucleus, energy would be given off, which would correspond to the increased stability of the product nucleus per nuclear particle.

SUBAREA III.   **ENERGY, CHEMICAL BONDS,
                                                    AND MOLECULAR STRUCTURE**

**COMPETENCY 0011   UNDERSTAND THE PRINCIPLES OF
                    THERMODYNAMICS AND CALORIMETRY.**

Kinetic molecular theory says that the particles of a substance are in constant random motion. What causes this motion? Energy.

Energy is the ability or capacity to do work or supply heat and heat is a form of energy, then heat must be able to do work. Work is force times distance. $W = Fd$

Heat energy is absorbed by atoms and molecules making up a substance that causes the atoms and molecules to move more. In the solid state, the motion is little more than a vibration; this vibration separates the atoms from each other, moving them apart, decreasing the force of attraction between the atoms or molecules, causing a phase change to liquid. As a liquid, the molecules still have attractive forces acting between them but the addition of more heat forces the atoms or molecules farther apart weakening the forces to the point where they can no longer keep the atoms or molecules together. The substance has entered the gas phase.

How is the heat energy of the molecule measured? No device will measure heat energy directly. However, a property called temperature can give us an idea of the heat energy of the atoms or molecules. Temperature does not measure the individual temperature of the molecules but rather how fast the molecules are moving (kinetic energy) which is tied to

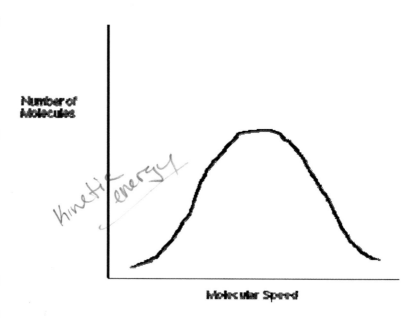

heat possessed by the molecules. A thermometer measures temperature.

A thermometer in a beaker of water will determine the 'temperature' of the water. Place the beaker on a hot plate and slowly allow the 'heat' to increase, the 'temperature' increases. Place the beaker in ice water, the 'temperature' decreases. Why does the thermometer respond this way?

As heat energy is added to a system, the matter absorbs the heat. This increased heat causes the atoms or molecules to move more (increase of kinetic energy). Each water molecule has a kinetic energy equal to $1/2\ m \cdot v^2$.

As the molecules move about they will bump into one another, the walls of the beaker and the thermometer. Some of the energy they possess is transferred to the bulb of the thermometer. The energy is transferred to the glass and, somehow, gets to the liquid inside the thermometer. The liquid is made of molecules. The increased energy of the molecules causes them to move faster. Moving faster they cause the liquid's volume to increase. With only one place to go the liquid moves up through the space in the stem. The thermometer can't measure the individual energy given by one molecule. It responds to all the molecules striking it. Some are moving faster than others. The observed temperature is the average kinetic energy of all the molecules striking the thermometer. The temperature of the water is a measure of the average kinetic energy of the water molecules involved in the process. How did the molecules get their kinetic energy? Heat.

Technically, heat is the measure of the internal energy gained or lost when two objects come in contact with one another and the temperature difference between the two objects is the measure of heat since it can not be directly determined.

The heat lost or gained is what is calculated.

$$\textit{Heat Loss} = \textit{Heat Gained}$$

$$m_{loss} \cdot c_{loss} \cdot \Delta T_{loss} = m_{gain} \cdot c_{gain} \cdot \Delta T_{gain}$$

where m= mass of substance undergoing temperature change,
c= heat capacity (specific heat) of the substance undergoing temperature change and ΔT = change in temperature.

If a time graph was made a pure substance being heated or cooled, it would look something like the following graph for the heating of water and be called a heating curve for water.

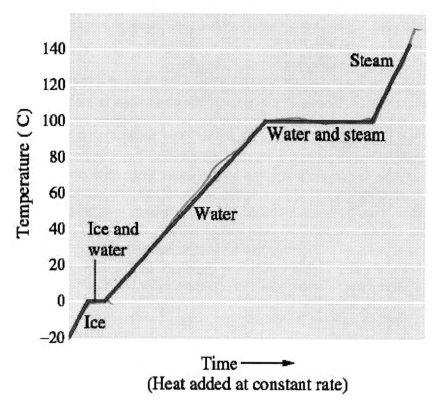

Time ⟶
(Heat added at constant rate)

Different changes are taking place during each interval on the graph.

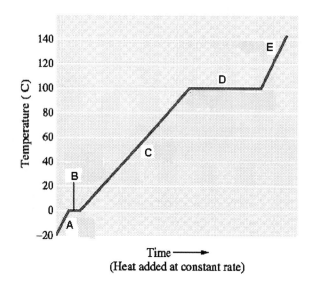

Time ⟶
(Heat added at constant rate)

When the system is heated, energy is transferred into it. In response to the energy it receives, the system changes, for example by increasing its temperature.

During the interval marked A on the graph, energy is being absorbed by the water molecules to increase the temperature to water's melting point, 0°C. The slope of the line for this interval shows the increase in temperature and is related to the heat capacity of the substance.

During the interval marked B on the graph, energy is still being added to the water but the temperature remains the same, at 0°C or water's melting point temperature. The additional energy is being used to overcome the intermolecular forces holding the water molecules in their solid pattern. This energy is moving the particles apart, breaking or weakening the forces of attraction trying to keep the water molecules aligned. The solid water (ice) is being converted to liquid water; a phase change is occurring. The temperature will not increase until every solid particle has melted and the entire sample is liquid.

Temperature again increases during interval C on the graph. Energy is being absorbed by the liquid water molecules. Notice that the slope of the line during this interval is different than the slope of the line during interval A. Again, this is due to differences in the heat capacity of ice and liquid water.

The flat line during interval D indicates that a phase change is occurring here. The additional energy is being used to overcome the attractive forces holding the liquid water molecules together. The water molecules increase their kinetic energies and move farther apart, changing to water vapor molecules. This occurs at the boiling point temperature, or 100°C in the case of water. The temperature stays at the boiling point temperature until all water molecules are converted to vapor molecules.

Once this conversion occurs, the temperature increases as energy is added, reflective of the heat capacity of the substance as a vapor.

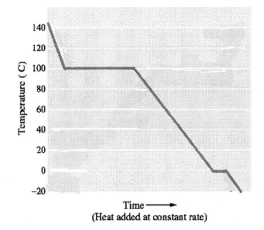

Time ⟶
(Heat added at constant rate)

Interpret this curve. Does it represent a heating or cooling curve?

Cooling

...ergy is the **driving force for change**. Energy has units of joules (J). ...mperature remains constant during phase changes, so the **speed** of ...olecules and their **translational kinetic energy do not change** during a change in phase.

The **internal energy** of a material is the **sum of the total kinetic energy** of its molecules and the **potential energy** of interactions between those molecules. Total kinetic energy includes the contributions from translational motion and other components of motion such as rotation. The potential energy includes **energy stored in the form of resisting intermolecular attractions** between molecules.

The **enthalpy** (*H*) of a material is the **sum of its internal energy and the mechanical work** it can do by driving a piston. We usually don't deal with mechanical work in high school chemistry, so the differences between internal energy and enthalpy are not important. The key concept is that a change in the **enthalpy** of a substance is the total **energy** change caused by **adding/removing heat** at constant pressure.

When a material is heated and experiences a phase change, **thermal energy is used to break the intermolecular bonds** holding the material together. Similarly, bonds are formed with the release of thermal energy when a material changes its phase during cooling. Therefore, **the energy of a material increases during a phase change that requires heat and decreases during a phase change that releases heat**. For example, the energy of H_2O increases when ice melts and decreases when water freezes.

Hess's law states that energy changes are state functions. The amount of energy depends only on the states of the reactants and the state of the products, but not on the intermediate steps. Energy (enthalpy) changes in chemical reactions are the same, regardless whether the reactions occur in one or several steps. The total energy change in a chemical reaction is the sum of the energy changes in its many steps leading to the overall reaction.

A **standard** thermodynamic value occurs with all components at 25 °C and 100 kPa. This *thermodynamic standard state* is slightly different from the *standard temperature and pressure* (STP) often used for gas law problems (0 °C and 1 atm=101.325 kPa). Standard properties of common chemicals are listed in tables.

The **heat of formation, ΔH_f** ,of a chemical is the heat required (positive) or emitted (negative) when elements react to form the chemical. It is also called the enthalpy of formation. The **standard heat of formation $\Delta H_f°$** is the heat of formation with all reactants and products at 25 °C and 100 kPa.

Elements in their **most stable form** are assigned a value of $\Delta H_f° = 0$ kJ/mol. Different forms of an element in the same phase of matter are known as **allotropes**.

Example: The heat of formation for carbon as a gas is:

$$\Delta H_f^\circ \text{ for } C(g) = 718.4 \ \frac{kJ}{mol}.$$

Carbon in the solid phase exists in three allotropes. A C_{60} *buckyball* (one face is shown to the left), contains C atoms linked with aromatic bonds and arranged in the shape of a soccer ball. C_{60} was discovered in 1985. *Diamond* (below left) contains single C–C bonds in a three dimensional network. The most stable form at 25 °C is *graphite* (below right). Graphite is composed of C atoms with aromatic bonds in sheets.

$$\Delta H_f^\circ \text{ for } C_{60}(\textit{buckminsterfullerene or buckyball}) = 38.0 \ \frac{kJ}{mol}$$

$$\Delta H_f^\circ \text{ for } C_\infty \ (\textit{diamond}) = 1.88 \ \frac{kJ}{mol}$$

$$\Delta H_f^\circ \text{ for } C_\infty \ (\textit{graphite}) = 0 \ \frac{kJ}{mol}.$$

Heat of combustion ΔH_c (also called enthalpy of combustion) is the heat of reaction when a chemical **burns in O_2** to form completely oxidized products such as **CO_2 and H_2O**. It is also the heat of reaction for **nutritional molecules that are metabolized** in the body. The standard heat of combustion ΔH_c° takes place at 25 °C and 100 kPa. **Combustion is always exothermic**, so the negative sign for values of ΔH_c is often omitted. If a combustion reaction is used in Hess's Law, the value must be negative.

Example: Determine the standard heat of formation ΔH_f° for ethylene:
$$2C(graphite) + 2H_2(g) \rightarrow C_2H_4(g).$$

Use the heat of combustion for ethylene:

$$\Delta H_c^\circ = 1411.2 \frac{kJ}{mol\ C_2H_4} \quad \text{for}\ \ C_2H_4(g) + 3O_2(g) \rightarrow 2CO_2(g) + 2H_2O(l)$$

(handwritten: reactant)

and the following two heats of formation for CO_2 and H_2O:

$$\Delta H_f^\circ = -393.5 \frac{kJ}{mol\ C} \quad \text{for}\quad C(graphite) + O_2(g) \rightarrow CO_2(g)$$

$$\Delta H_f^\circ = -285.9 \frac{kJ}{mol\ H_2} \quad \text{for}\quad H_2(g) + \frac{1}{2}O_2(g) \rightarrow H_2O(l).$$

Solution: Use Hess's Law after rearranging the given reactions so they cancel to yield the reaction of interest. Combustion is exothermic, so ΔH for this reaction is negative. We are interested in C_2H_4 as a product, so we take the opposite (endothermic) reaction. The given ΔH are multiplied by stoichiometric coefficients to give the reaction of interest as the sum of the three:

(handwritten: product)

$$2CO_2(g) + 2H_2O(l) \rightarrow C_2H_4(g) + 3O_2(g) \qquad \Delta H = 1411.2 \frac{kJ}{mol\ reaction}$$

(handwritten: Flipped)

$$2C(graphite) + 2O_2(g) \rightarrow 2CO_2(g) \qquad \Delta H = -787.0 \frac{kJ}{mol\ reaction}$$

$$2H_2(g) + O_2(g) \rightarrow 2H_2O(l) \qquad \Delta H = -571.8 \frac{kJ}{mol\ reaction}$$

(handwritten: multiply by coefficient) (-393.5)(2) (-285.9)(2))

$$2C(graphite) + 2H_2(g) \rightarrow C_2H_4(g) \qquad \Delta H_f^\circ = 52.4 \frac{kJ}{mol}$$

(handwritten: - cancel out that elements are same on each side)

For example, in the diagram below, we look at the oxidation of carbon into CO and CO_2. The direct oxidation of carbon (graphite) into CO_2 yields an enthalpy of -393 kJ/mol. When carbon is oxidized into CO and then CO is oxidized to CO_2, the enthalpies are -110 and -283 kJ/mol respectively. The sum of enthalpy in the two steps is exactly -393 kJ/mol, same as the one-step reaction.

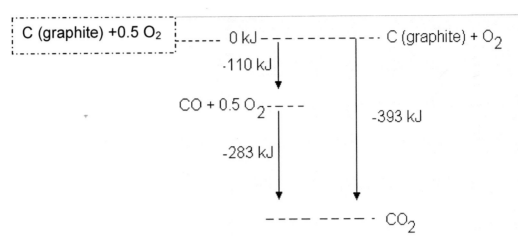

The two-step reactions are:

$$C + \tfrac{1}{2} O_2 \rightarrow CO, \qquad \Delta H° = -110 \text{ kJ/mol}$$
$$CO + \tfrac{1}{2} O_2 \rightarrow CO_2, \qquad \Delta H° = -283 \text{ kJ/mol}.$$

Adding the two equations together and cancel out the intermediate, CO, on both sides leads to

$$C + O_2 \rightarrow CO_2, \qquad \Delta H° = (-110)+(-283) = -393 \text{ kJ/mol}.$$

Application of Hess's law enables us to calculate ΔH, $\Delta H°$, and ΔH_f for chemical reactions that impossible to measure, providing that we have all the data of related reactions.

For example:

The enthalpy of combustion for H_2, C(graphite) and CH_4 are -285.8, -393.5, and -890.4 kJ/mol respectively. Calculate the standard enthalpy of formation ΔH_f for CH_4.

Using the equations and their ΔH values
(1) $H_2(g) + 0.5\ O_2(g) \rightarrow H_2O(l)$ $\Delta H = -285.8$ kJ/mol
(2) $C(graphite) + O_2(g) \rightarrow CO_2(g)$ $\Delta H = -393.5$ kJ/mol
(3) $CH_4(g) + 2O_2(g) \rightarrow CO_2(g) + 2H_2O(l)$ $\Delta H = -890.4$ kJ/mol

Find: $C + 2H_2 \rightarrow CH_4$

2 $(H_2(g) + 0.5\ O_2(g) \rightarrow H_2O(l))$ $\Delta H = -285.8$ kJ/mol) × 2

or $2H_2(g) + O_2(g) \rightarrow 2H_2O(l)$ $\Delta H = -571.6$ kJ

+ $C(graphite) + O_2(g) \rightarrow CO_2(g)$ $\Delta H = -393.5$ kJ

+ $CO_2(g) + 2H_2O(l) \rightarrow CH_4(g) + 2O_2(g)$ $\Delta H = +890.4$ kJ   opposite reaction now endothermic

Flip reaction → to have CH_4 as a product

———————————————————————————————————————

$C(graphite) + 2H_2(g) \rightarrow CH_4(g)$ $\Delta H = -74.7$ kJ

From these data, we can construct an energy level diagram for these chemical combinations as follows:

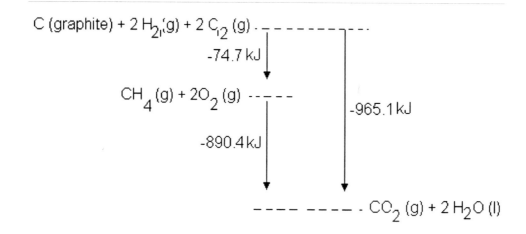

COMPETENCY 0012 UNDERSTAND ENERGY RELATIONSHIPS IN CHEMICAL BONDING AND CHEMICAL REACTIONS.

Chemical energy is an important source of energy. It is the energy stored in substances by virtue of the arrangement of atoms within the substance. When atoms are rearranged during chemical reactions, energy is either released or consumed. It is the energy released from chemical reactions that fuel our economy and power our bodies. Most of the electricity produced comes from chemical energy released by the burning of petroleum, coal and natural gas. ATP is the molecule used by our body to carry chemical energy form cell to cell.

The energy found in molecules is found in the bonds between the atoms in the molecule. To break these bonds requires energy. Once broken apart, the atoms, ions or molecules rearrange themselves to form new substances, making new bonds. Making new bonds releases energy.

If during a chemical reaction, more energy is needed to break the reactant bonds than released when product bonds form, the reaction is endothermic and heat is absorbed from the environment becomes colder.

On the other hand, if more energy is released due to product bonds forming than is needed to break reactant bonds the energy is exothermic and the excess energy is released to the environment as heat. The temperature of the environment goes up.

The total energy absorbed or released in the reaction can be determined by using bond energies. The total energy change of the reaction is equal to the total energy of all of the bonds of the products minus the total energy of all of the bonds of the reactants.

For example, propane , C_3H_8, is a common fuel used in heating homes and backyard grills. When burned, excess energy from the combustion reaction is released and used to cook our food, for example.

$$C_3H_8 \text{ (g)} + 5\,O_2 \text{ (g)} \rightarrow 3\,CO_2 \text{ (g)} + 4\,H_2O(l)$$

The total energy of the products is from the bonds found in the carbon dioxide molecules and the water molecules.

$$3\ O=C=O\ +\ 4\ H\!\diagup\!\!\overset{O}{}\!\!\diagdown\!H$$

or 6 C=O bonds and 8 H-O bonds.

A table of bond energies gives the following information:

C=O   743 kJ/mol

H-O   463 kJ/mol

So for these molecules there would be: (6 x 743 kJ) + 8 x 463 kJ) =8162 kJ of energy released when these molecules form.

The reactants are these:

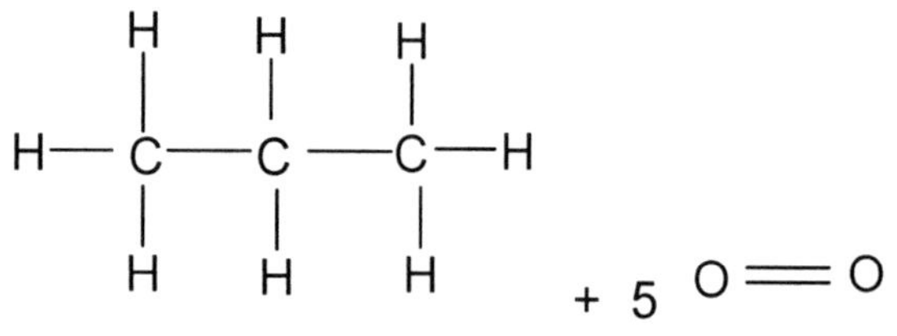

$+ 5 \ O = O$

or

2 C-C bonds
8 C-H bonds and
5 O=O bonds

These bonds require the following energy to break:

C-C   348 kJ/ mol
C-H   412 kJ/mol
O=O 498 kJ.mol

The total energy required for the reactants would be

     (2 x 348 kJ) + (8 x 412 kJ) + (5 x 498 kJ) = 6482 kJ of energy.

The total energy change that occurs during the combustion of propane is then:

8162 kJ – 6482 kJ = 680 kJ of energy is released for every mole of propane that burns in excess oxygen.

COMPETENCY 0013   UNDERSTAND THE TYPES OF BONDS BETWEEN ATOMS (INCLUDING IONIC, COVALENT AND METALLIC BONDS), THE FORMATION OF THESE BONDS, AND THE PROPERTIES OF SUBSTANCES CONTAINING THE DIFFERENT BONDS.

Chemical compounds form when two or more atoms join together. A stable compound occurs when the total energy of the combination of the atoms together is lower than the atoms separately. The combined state suggests an attractive force exists between the atoms. This attractive force is called a chemical bond.

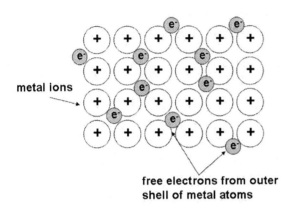

metal ions

free electrons from outer shell of metal atoms

Metallic bonds occur when the bonding is between two metals. Metallic properties, such as low ionization energies, conductivity, and malleability, suggest that metals possess strong forces of attraction between atoms, but still have electrons that are able to move freely in all directions throughout the metal. This creates a "sea of electrons" model where electrons are quickly and easily transferred between metal atoms. In this model, the outer shell electrons are free to move. The metallic bond is the force of attraction that results from the moving electrons and the positive nuclei left behind. The strength of metal bonds usually results in regular structures and high melting and boiling points.

Ionic Bonds
An **ionic bond** describes the electrostatic forces that exist between **particles of opposite charge.** An ionic bond is the result of an atom(s) losing electrons to form a positive ion being attracted to an atom(s) that has gained electrons to form a negative ion.

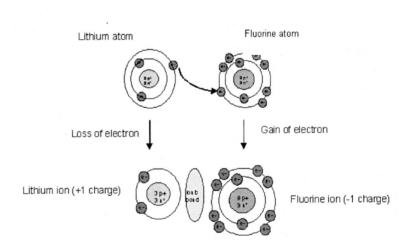

Lithium atom

Fluorine atom

Loss of electron

Gain of electron

Lithium ion (+1 charge)

Fluorine ion (-1 charge)

Due to low ionization energies, metals have a tendency to lose valence electrons relatively easily, whereas non-metals, which have high ionization energies and high electronegativites, gain electrons easily.

This produces cations (positively charged) and anions (negatively charged). Coulomb's Law says that opposites attract, and so do the oppositely charged ions. This *electrostatic interaction* between the anion and the cation results in an ionic bond. Ionic bonds will only occur when a metal is bonding to a non-metal, and is the result of the periodic trends (ionization energy and electronegativity). Elements that form an ionic bond with each other have a large difference in their electronegativity.

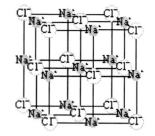

Anions and cations pack together into a crystal **lattice** as shown to the right for NaCl. Ionic compounds are also known as **salts**.

Single and multiple covalent bonds

A **covalent bond** forms when at least one pair of electrons is shared by two atoms. The shared electrons are found in the valence energy level and lead to a lower energy if they are shared in a way that creates a noble gas configuration (a full octet). Covalent, or molecular, bonds occur when a non-metal is bonding to a non-metal. This is due primarily to the fact that non-metals have high ionization energies and high electronegativities. Neither atom wants to give up electrons; both want to gain them. In order to satisfy both octets, the electrons will be shared between the two atoms.

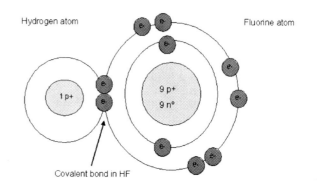

Hydrogen atom

Fluorine atom

1 p+

9 p+
9 n°

Covalent bond in HF

Sharing of electrons can be equal or unequal, resulting in a separation of charge (polar) or an even distribution of charge (non-polar). The polarity of a bond can be determined through an examination of the electronegativities of the atoms involved in the bond. The more electronegative atom will have a stronger attraction to the electrons, thus possessing the electrons more of the time. This results in a partial negative charge (δ^-) on the more electronegative atom and a partial positive charge (δ^+) on the less electronegative atom.

The simplest covalent bond is between the two single electrons of hydrogen atoms. Covalent bonds may be represented by an electron pair (a pair of dots) or a line as shown below. The shared pair of electrons provides each H atom with two electrons in its valence shell (the $1s$ orbital), so both have the stable electron configuration of helium.

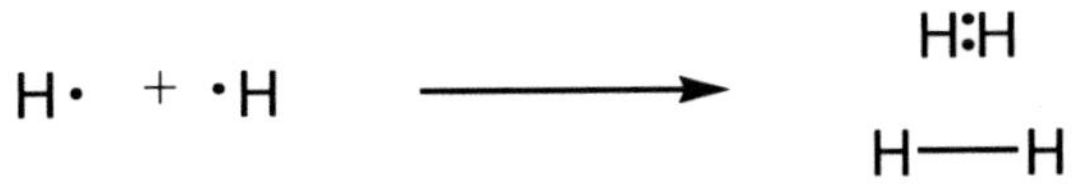

H· + ·H $\longrightarrow$ H:H

H——H

Chlorine molecules have 7 electrons in their valence shell and share a pair of electrons so both have the stable electron configuration of argon.

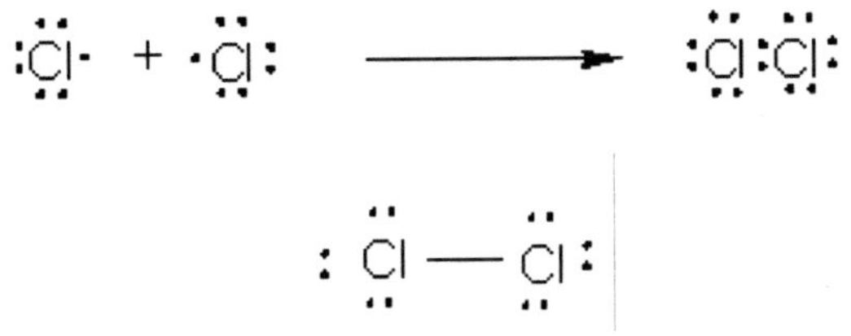

:Cl· + ·Cl: $\longrightarrow$:Cl:Cl:

:Cl——Cl:

In the previous two examples, a single pair of electrons was shared, and the resulting bond is referred to as a **single bond**. When two electron pairs are shared, two lines are drawn, representing a **double bond**, and three shared pairs of electrons represents a **triple bond** as shown below for CO_2 and N_2. The remaining electrons are in **unshared pairs**.

O::C::O

O═C═O

:N::N:

:N≡≡N:

Network solids

It is possible for many atoms to link up to form a giant covalent structure or lattice. The atoms are usually non-metals. This produces a very strong 3-dimensional covalent bond network or lattice. This gives them significantly different properties from the small simple covalent molecules. This is illustrated by carbon in the form of diamond (an allotrope of carbon). Carbon has four outer electrons that form four single bonds, so each carbon bonds to four others by electron pairing/sharing. Pure silicon, another element in Group 4, has a similar structure.

diamond

- ○ NOTE: Allotropes are different forms of the same element in the same physical state. They occur due to different bonding arrangements and so diamond, graphite and fullerenes are the three solid allotropes of the element carbon.

  Oxygen (dioxygen), O_2, and ozone (trioxygen), O_3, are the two small gaseous allotrope molecules of the element oxygen.

  - ▪ Sulfur has three solid allotropes, two different crystalline forms based on small S_8 molecules called rhombic and monoclinic sulfur and a 3rd form of long chain (-S-S-S- etc.) molecules called plastic sulfur.

This type of giant covalent structure is thermally very stable and has a very high melting and boiling points because of the strong covalent bond network (3D or 2D in the case of graphite). They are usually poor conductors of electricity because the electrons are not usually free to move as they can in metallic structures.

Also because of the strength of the bonding in all directions in the structure, they are often very strong and will not dissolve in solvents like water. The bonding network is too strong to allow the atoms to become surrounded by solvent molecules

graphite

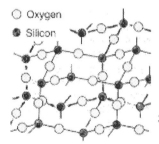

Silicon dioxide (silica, SiO_2) has a similar 3D structure and properties. The hardness of diamond enables it to be used as the 'leading edge' on cutting tools.

○ Oxygen
● Silicon

Silicon dioxide

Polar/nonpolar covalent bonds

Electron pairs shared between **two atoms of the same element are shared equally**. At the other extreme, **for ionic bonding there is no electron sharing** because the electron is transferred completely from one atom to the other. Most bonds fall somewhere between these two extremes, and the electrons are **shared unequally**. This will increase the probability that the shared electrons will be located on one of the two atoms, giving that atom a **partial negative charge**, and the other atom a **partial positive charge** as shown below for gaseous HCl. Such bonds are referred to as **polar bonds**. A particle with a positive and a negative region is called a **dipole**.

A lower-case delta (δ) is used to indicate partial charge or an arrow is draw from the partial positive to the partial negative atom.

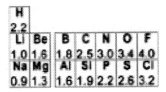

Electronegativity is a measure of **the ability of an atom to attract electrons** in a chemical bond. Metallic elements have low electronegativities and nonmetallic elements have high electronegativities .

Linus Pauling developed the concept of electronegativity and its relationship to different types of bonds in the 1930s.

A **large electronegativity difference** (greater than 1.7) results in an **ionic bond**. Any bond composed of two different atoms will be slightly polar, but for a **small electronegativity difference** (less than 0.4), the distribution of charge in the bond is so nearly equal that the result is called a **nonpolar covalent bond**. An **intermediate electronegativity difference** (from 0.4 to 1.7) results in a **polar covalent bond**. HCl is polar covalent because Cl has a very high electronegativity (it is near F in the periodic table) and H is a nonmetal (and so it will form a covalent bond with Cl), but H is near the dividing line between metals and nonmetals, so there is still a significant electronegativity difference between H and Cl. Using the numbers in the table above, the electronegativity for Cl is 3.2 and it is 2.2 for H. The difference of 3.2 − 2.2 = 1.0 places this bond in the middle of the range for polar covalent bonds.

Bond type is actually a continuum as shown in the following chart for common bonds. Note that the **C-H bond** is considered **nonpolar**.

Type of bonding	Electronegativity difference	Bond
		Fr^+-F^-
Very ionic		Na^+-F^-
	3.0	
⋮	⋮	⋮
Ionic		Na^+-Cl^-
	2.0	Na^+-Br^-
Mostly ionic		Na^+-I^-
Mostly polar covalent	1.5	C^+-F^-
		H^+-O^-
Polar covalent	1.0	H^+-Cl^-
		$C^{\delta+}=O^{\delta-}$
		H^+-N^-
		C^+-Cl^-
	0.5	$C^{\delta+}\equiv N^{\delta-}$
Mostly nonpolar covalent		$C-H$
Fully nonpolar covalent	0	$H_2, N_2, O_2,$ $F_2, Cl_2, Br_2, I_2,$ $C-C, S-S$

Increasing ionic character ⇑

Physical properties of substances with different bonds

To a large degree, the physical properties of a solid substance are determined by the type of bonding which holds the molecules, atoms or ions together in a solid. Depending on the type of bonding, solids may be described as ionic, molecular, metallic or covalent network solids. Each type of bonding affects the physical properties of a solid in different ways. The type of bonding in a solid can be determined by studying its physical properties.

Ionic solids contain ions held together by ionic bonds. These solids are typically hard and have high melting points. They are often soluble in water, but insoluble in organic solvents. They do not conduct electricity in the solid state, but do so both in aqueous solutions and in pure molten form. (Use a solid like calcium sulfate as an ionic solid, which is not soluble in water.)

Molecular solids are made up of molecules held together by relatively weak forces. They tend to be soft, easily melted and volatile. Molecular solids are likely to be insoluble in water, but soluble in organic solvents. Neither the solid nor the molten state conducts electricity, nor do any of the solutions. (Use solid sucrose as a molecular solid, which does dissolve in water.)

Covalent network solids contain atoms held together by a network of covalent bonds that link every atom in the solid to every other atom. The molecules are gigantic; each particle of the crystal is essentially one molecule. This type of solid is hard, nonvolatile, with a very high melting point and insoluble in both water and inorganic solvents. They do not conduct electricity.

Metallic solids contain atoms bonded together by metallic bonds. These bonds are strong but not localized. Since the electrons in the metallic bonds are relatively mobile, metals tend to have high melting points and be hard, malleable, nonvolatile and shiny. Metals are soluble neither in water nor organic solvents. Some metals, such as sodium, dissolve by reacting with water. Metals sometimes dissolve in liquid metallic mercury.

The chart below summarizes the physical properties of the various types of solid substances.

Physical Properties of Substances				
Property	Observations			
	NaCl (ionic)	$C_6H_4Cl_2$ (molecular)	SiO_2 (covalent)	Fe (metallic)
Hardness	Brittle	Soft	Hard	hard
Volatility/odor	None	Mothballs	None	none
Melting	Just melts	Easily melts	Doesn't melt	Doesn't melt
Water solubility	Soluble	Insoluble	Insoluble	insoluble
Cyclohexane solubility	insoluble	Soluble	Insoluble	insoluble
Solid conductivity	No	No	No	Yes
Water solution conductivity	Yes	No	No	No
Cyclohexane solution conductivity	No	No	No	No

COMPETENCY 0014 UNDERSTAND TYPES AND CHARACTERISTICS OF MOLECULAR INTERACTION AND PROPERTIES OF SUBSTANCES CONTAINING DIFFERENT TYPES OF INTERACTIVE FORCES BETWEEN MOLECULES.

Polarity is a physical property of compounds which relates other physical properties such as melting and boiling points, solubility, and intermolecular interactions between molecules.

For the most part, there is a direct correlation between the polarity of a molecule and number and types of polar or non-polar covalent bonds which are present.

In a few cases, a molecule may have polar bonds, but in a symmetrical arrangement which then gives rise to a non-polar molecule such as carbon dioxide.

Polar and nonpolar molecules
A **polar molecule** has positive and negative regions as shown below for HCl.

$$\delta + \qquad \delta -$$
$$H\!\!-\!\!Cl \qquad \qquad H\!\!-\!\!Cl \longrightarrow$$

Bond polarity is **necessary but not sufficient for molecular polarity**. A molecule containing polar bonds will still be nonpolar if the most negative and most positive location occurs at the same point. In other words, **in a polar molecule, bond polarities must not cancel**.

To determine if a molecule is polar perform the following steps.

1) Draw the molecular structure.
2) Assign a polarity to each bond with an arrow (remember C-H is nonpolar). If none of the bonds is polar, the molecule is nonpolar.
3) Determine if the polarities cancel each other in space. If they do, the molecule is nonpolar. Otherwise the molecule is polar.

Examples: Which of the following are polar molecules: CO_2, CH_2Cl_2, CCl_4.

Solution: 1)

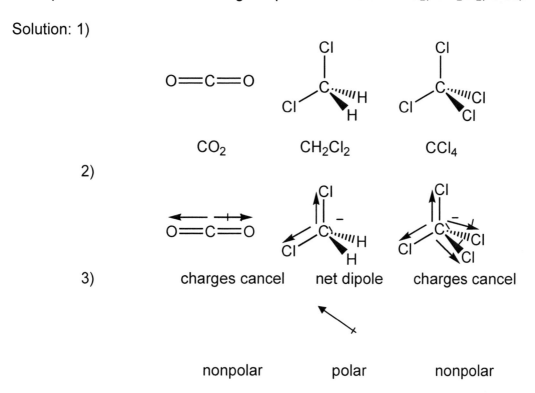

2)

3)

The polarity of molecules is critical for determining a good solvent for a given solute. Additional practice on the topic of polar bonds and molecules is available at http://cowtownproductions.com/cowtown/genchem/09_17M.htmu

The polarity of molecules also creates attractive forces between molecules that cause the molecules to "stick" together. These attractive forces are called **Intermolecular Forces.** The physical properties of melting point, boiling point, vapor pressure, evaporation, viscosity, surface tension, and solubility are related to the strength of attractive forces between molecules.

There are four types of intermolecular forces. Most of the intermolecular forces are identical to bonding between atoms in a single molecule. Intermolecular forces just extend the thinking to forces **between** molecules and follows the patterns already set by the bonding within molecules.

The following intermolecular forces between molecules are usually weaker than covalent, ionic, and metallic bonds. They are listed from the strongest force to the weakest.

Ionic Forces

The forces holding ions together in ionic solids are electrostatic forces. Opposite charges attract each other. These are the strongest intermolecular forces. Ionic forces hold many ions in a crystal lattice structure

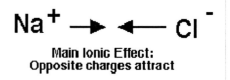

Main Ionic Effect:
Opposite charges attract

Ion-dipole interactions

Salts tend to dissolve in several polar solvents. An ion with a full charge in a polar solvent will **orient nearby solvent molecules** so that their opposite partial charges are pointing towards the ion. In aqueous solution, certain salts react to form solid **precipitates** if a combination of their ions is insoluble.

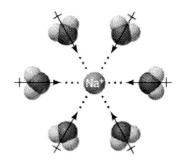

Ion-dipole

Hydrogen Bonding In Water

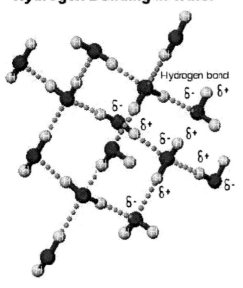

Hydrogen bond

Hydrogen bonds

Hydrogen bonds are particularly **strong dipole-dipole interactions** that form between the H-atom of one molecule and an **F, O, or N** atom of an adjacent molecule. The partial positive charge on the hydrogen atom is attracted to the partial negative charge on the electron pair of the other atom. The hydrogen bond between two water molecules is shown as the dashed line below:

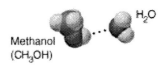

Methanol
(CH_3OH)

H_2O

H bond

Dipole-dipole interactions

The intermolecular forces between polar molecules are known as dipole-dipole interactions. The partial positive charge of one molecule is attracted to the partial negative charge of its neighbor.

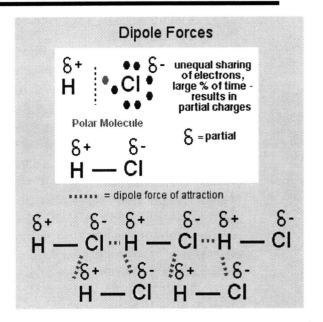

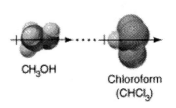

Dipole-dipole

Ion-induced dipole

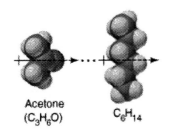

When a nonpolar molecule (or a noble gas atom) encounters an ion, its **electron density is temporarily distorted** resulting in an **induced dipole** that will be attracted to the ion. Intermolecular attractions due to induced dipoles in a nonpolar molecule are known as **London forces or Van der Waals interactions**. These are very weak intermolecular forces.

Ion-induced dipole

For example, carbon tetrachloride, CCl_4, has polar bonds but is a nonpolar molecule due to the symmetry of those bonds. An aluminum cation will draw the unbonded electrons of the chlorine atom towards it, distorting the molecule and creating an attractive force as shown by the dashed line above.

Dipole-induced dipole

The partial charge of **a permanent dipole may also induce a dipole in a nonpolar molecule** resulting in an attraction similar to—but weaker than—that created by an ion.

Acetone
(C_3H_6O) C_6H_{14}

Dipole-induced dipole

London dispersion force: induced dipole-induced dipole
The above two examples required a permanent charge to induce a dipole in a nonpolar molecule. A nonpolar molecule may also induce a temporary dipole on its identical neighbor in a pure substance. These forces occur because at any given moment, electrons are located within a certain region of the molecule, and **the instantaneous location of electrons will induce a temporary dipole** on neighboring molecules. For example, an isolated helium atom consists of a nucleus with a 2+ charge and two electrons in a spherical electron density cloud. An attraction of He atoms due to London dispersion forces (shown at right by the dashed line) occurs because when the electrons happen to be distributed unevenly on one atom, a dipole is induced on its neighbor. This dipole is due to intermolecular repulsion of electrons and the attraction of electrons to neighboring nuclei.

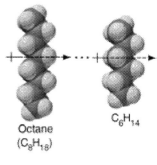

Octane
(C_8H_{18})

C_6H_{14}

Dispersion

The strength of London dispersion forces **increases for larger molecules** because a larger electron cloud is more easily polarized. The strength of London dispersion forces also **increases for molecules with a larger surface area** because there is greater opportunity for electrons to influence neighboring molecules if there is more potential contact between the molecules. Paraffin in candles is an example of a solid held together by weak London forces between large molecules. These materials are soft.

Impact on physical properties
The impact of intermolecular forces on substances is best understood by imagining ourselves shrinking down to the size of molecules and picturing what happens when we stick more strongly to molecules nearby. It will take more energy (higher temperatures) to pull us away from our neighbors.

If two substances are being compared, the material with the **greater intermolecular attractive forces** (i.e. the stronger intermolecular bond) will have the following properties relative to the other substance:

For solids:
Higher melting point
Higher enthalpy of fusion
Greater hardness
Lower vapor pressure

For liquids:
Higher boiling point
Higher critical temperature
Higher critical pressure
Higher enthalpy of vaporization
Higher viscosity
Higher surface tension
Lower vapor pressure

For gases:
Intermolecular attractive forces are neglected for ideal gases.

For example, covalent networks, salts, and metals are nearly always solids at room temperature because the strength of these bonds results in a high melting point. H_2O and NH_3 are liquids at room temperature because they contain hydrogen bonds. These bonds are of intermediate strength, so the melting point of these compounds is lower than room temperature and their boiling point is higher than room temperature. H_2S contains weaker dipole-dipole interactions than H_2O because the sulfur atoms do not form hydrogen bonds. Therefore, H_2S is a gas at room temperature due to its low boiling point. Finally, small non-polar molecules such as CO_2 (44.01 u), N_2 or atoms such as He will be gases at room temperature due to very weak London forces, but larger non-polar molecules such as octane (114.22 u) or CCl_4 (153.82 u) may be liquids, and very large non-polar molecules such as paraffins will be soft solids.

Properties of substances and their attractive forces are summarized in the chart below.

Properties of Substances				
Type of substance	Particles	Attractive Forces	Properties	Examples
Ionic	Ionic crystal	Electrostatic forces of attraction between positve ion and negative ion	MP -300°C-1000°C All solids Hard and brittle All soluble in water Good conductors in molten and aqueous states but not as solids	NaCl $MgBr_2$ BaO KNO_3
Molecular	Nonpolar molecules	Van der Waals forces	Very low M.P and B.P. Usually decomposes before boiling Not soluble in water Found as gases or liquids Non conductors (few are very poor conductors)	H_2 Cl_2 CCl_4
Molecular	Polar molecules	Dipole-dipole forces	Low M.P. and B.P. Usually liquids Poor conductors Soluble in water	H_2O HCl SO_2
Metallic	Metals	Metallic forces called sea of mobile electrons	M.P. 0-2000°C Not soluble in water Hard and soft metals Good conductors as solid and melted	Al Na Cu Au

COMPETENCY 0015 UNDERSTAND THE NOMENCLATURE AND STRUCTURE OF ORGANIC COMPOUNDS.

The IUPAC is the **International Union of Pure and Applied Chemistry**. **Organic compounds contain carbon,** and have their own branch of chemistry because of the huge number of carbon compounds in nature, including nearly all the molecules in living things. The 1979 IUPAC organic nomenclature is used and taught most often today, and it is the nomenclature described here.

The simplest organic compounds are called **hydrocarbons** because they **contain only carbon and hydrogen**. Hydrocarbon molecules may be divided into the classes of **cyclic** and **open-chain** depending on whether they contain a ring of carbon atoms. Open-chain molecules may be divided into **branched** or **straight-chain** categories.

Hydrocarbons are also divided into classes called **aliphatic** and **aromatic**. Aromatic hydrocarbons are related to benzene and are always cyclic. Aliphatic hydrocarbons may be open-chain or cyclic. Aliphatic cyclic hydrocarbons are called **alicyclic**. Aliphatic hydrocarbons are one of three types: alkanes, alkenes, and alkynes.

Alkanes
Alkanes contain only single bonds. Alkanes have the maximum number of hydrogen atoms possible for their carbon backbone, so they are called **saturated**. Alkenes, alkynes, and aromatics are **unsaturated** because they have fewer hydrogen atoms..

Straight-chain alkanes are also called **normal alkanes**. These are the simplest hydrocarbons. They consist of a linear chain of carbon atoms. The names of these molecules contain the suffix –*ane* and a **root based on the number of carbons in the chain** according to the table on the following page. The first four roots, *meth–*, *eth–*, *prop–*, and *but–* have historical origins in chemistry, and the remaining alkanes contain common Greek number prefixes. Alkanes have the general formula C_nH_{2n+2}.

A single molecule may be represented in multiple ways. Methane and ethane in the table are shown as three-dimensional structures with dashed wedge shapes attaching atoms behind the page and thick wedge shapes attaching atoms in front of the page.

Number of carbons	Name	Formula	Structure
1	Methane	CH_4	
2	Ethane	C_2H_6	
3	Propane	C_3H_8	
4	Butane	C_4H_{10}	
5	Pentane	C_5H_{12}	
6	Hexane	C_6H_{14}	
7	Heptane	C_7H_{16}	
8	Octane	C_8H_{18}	

Additional ways that pentane might be represented are:

n-pentane (the *n* represents a *normal* alkane)
$CH_3CH_2CH_2CH_2CH_3$
$CH_3(CH_2)_3CH_3$

If one hydrogen is removed from an alkane, the residue is called an **alkyl** group. The *–ane* suffix is replaced by an *–yl–* infix when this residue is used as **functional group**. Functional groups are used to systematically build up the names of organic molecules.

Branched alkanes are named using a four-step process:

1) Find the longest continuous carbon chain. This is the parent hydrocarbon.
2) Number the atoms on this chain beginning at the end near the first branch point, so the lowest locant numbers are used. Number functional groups from the attachment point.
3) Determine the numbered locations and names of the substituted alkyl groups. Use *di–*, *tri–*, and similar prefixes for alkyl groups represented more than once. Separate numbers by commas and groups by dashes.
4) List the locations and names of alkyl groups in alphabetical order by their name (ignoring the *di–*, *tri–* prefixes) and end the name with the parent hydrocarbon.

Example—Name the following hydrocarbon:

$$CH_2$$

$$H_3C-CH$$

$$CH-CH_3$$

$$H_2C-CH \qquad CH_3$$

$$H_3C \qquad H_2C-CH_2$$

Solution—
1) The longest chain is seven carbons in length, as shown by the bold lines below. This molecule is a heptane.
2) The atoms are numbered from the end nearest the first branch as shown:

$$CH_3^{1}$$

$$H_3C-CH^{2}$$

$$CH-CH_3$$
^{3}

$$H_2C-CH^{4} \qquad CH_3^{7}$$

$$H_3C \qquad H_2C=CH_2$$
$$\qquad\qquad 5 \qquad 6$$

3) Methyl groups are located at carbons 2 and 3 (2,3-dimethyl), and an ethyl group is located at carbon 4.
4) "Ethyl" precedes "methyl" alphabetically. The hydrocarbon name is: 4-ethyl-2,3-dimethylheptane.

The following branched alkanes have IUPAC-accepted common names:

Structure	Systematic name	Common name
	2-methylpropane	isobutane
	2-methylbutane	isopentane
	2,2-dimethylpropane	neopentane

The following alkyl groups have IUPAC-accepted common names. The systematic names assign a locant number of 1 to the attachment point:

Structure	Systematic name	Common name
	1-methylethyl	isopropyl
	2-methylpropyl	isobutyl
	1-methylpropyl	sec-butyl
	1,1-dimethylethyl	tert-butyl

Alkenes

Alkenes contain one or more double bonds. Alkenes are also called olefins. The suffix used in the naming of alkenes is –*ene*, and the number roots are those used for alkanes of the same length.

A number preceding the name shows the location of the double bond for alkenes of length four and above. Alkenes with one double bond have the general formula C_nH_{2n}. Multiple double bonds are named using –*diene*, –*triene*, etc. The suffix –*enyl*– is used for functional groups after a hydrogen is removed from an alkene. Ethene and propene have the common names **ethylene** and **propylene**. The ethenyl group has the common name **vinyl** and the 2-propenyl group has the common name **allyl**.

Examples:

H_2C=CH_2 is ethylene or ethene. H_2C=CH is a vinyl or ethenyl group.

is propylene or propene.   is an allyl or 2-propenyl group.

is 2-hexene.

is 2-methyl-1,3-butadiene (common name: isoprene).

Cis-trans isomerism is often part of the complete name for an alkene. Note that isoprene contains two adjacent double bonds, so it is a **conjugated** molecule.

Alkynes and alkenynes

Alkynes contain one or more triple bonds. They are named in a similar way to alkenes. The suffix used for alkynes is –*yne*. Ethyne is often called **acetylene**. Alkynes with one triple bond have the general formula C_nH_{2n-2}. Multiple triple bonds are named using –*diyne*, –*triyne*, etc. The infix –*ynyl*– is used for functional groups composed of alkynes after the removal of a hydrogen atom.

Hydrocarbons with **both double and triple bonds are known as alkenynes**. The locant number for the double bond precedes the name, and the locant for the triple bond follows the suffix –*en*– and precedes the suffix –*yne*.

Examples: $HC\equiv CH$ is acetylene or ethyne.

is 1-butyne.

$HC\equiv C-C\equiv C-CH_3$ is 1,3-pentadiyne.

is a 4-hexynyl group.

is 1-buten-3-yne. This compound has the common name of vinylacetylene

Cycloalkanes, –enes, and –ynes

Alicyclic hydrocarbons use the prefix cyclo– before the number root for the molecule. The structures for these molecules are often written as if the molecule lay entirely within the plane of the paper even though in reality, these rings dip above and below a single plane. When there is more than one substitution on the ring, numbering begins with the first substitution listed in alphabetical order.

Examples:

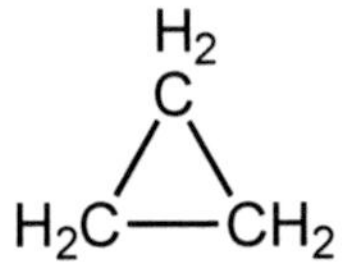

is cyclopropane

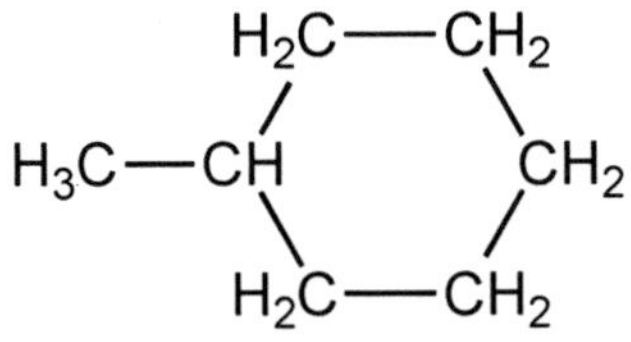

or

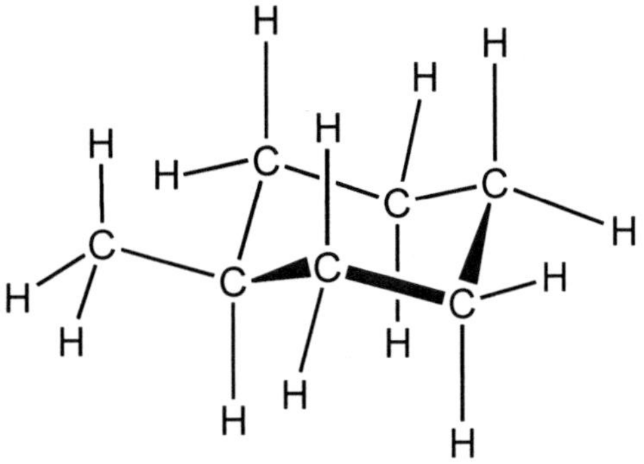

is methylcyclohexane.

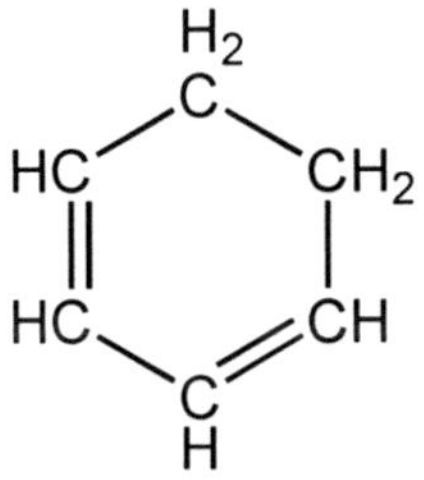

or

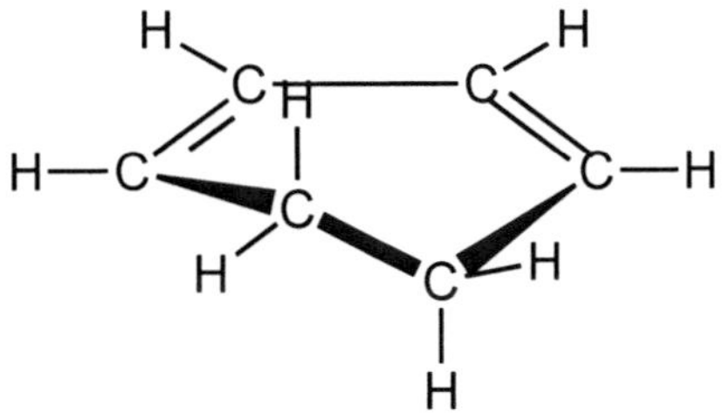

is 1,3-cyclohexadiene.

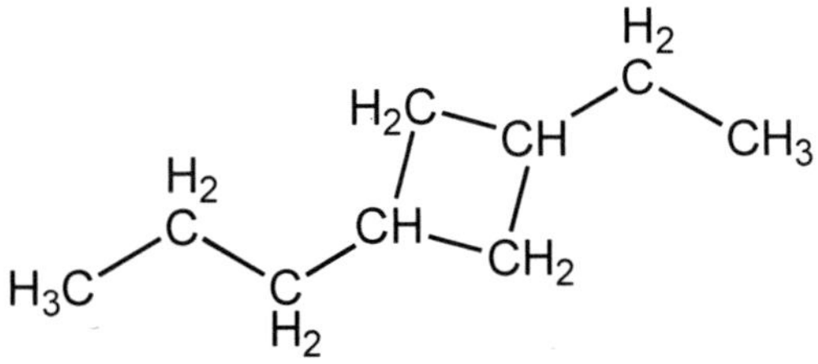

is 1-ethyl-3-propylcyclobutane.

Cis-trans isomerism is often part of the complete name for a cycloalkane.

Aromatic hydrocarbons

Aromatic hydrocarbons are structurally related to benzene or made up of benzene molecules fused together. These molecules are called **arenes** to distinguish them from alkanes, alkenes, and alkynes. All atoms in arenes lie in the same plane. In other words, aromatic hydrocarbons are flat. Aromatic molecules have electrons in delocalized π orbitals that are free to migrate throughout the molecule.

Substitutions onto the benzene ring are named in alphabetical order using the lowest possible locant numbers. The prefix *phenyl*– may be used for C_6H_5– (benzene less a hydrogen) attached as a functional group to a larger hydrocarbon residue. Arenes in general form aryl functional groups. A phenyl group may be represented in a structure by the symbol Ø. The prefix *benzyl*– may be used for $C_6H_5CH_2$– (methylbenzene with a hydrogen removed from the methyl group) attached as a functional group.

Examples:

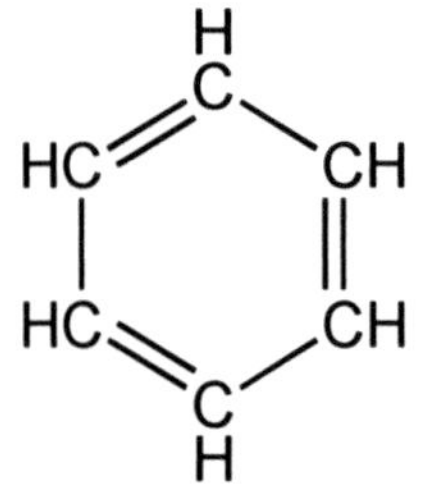

or

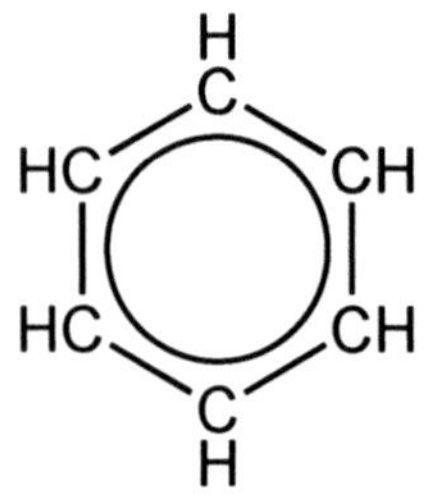

is benzene.

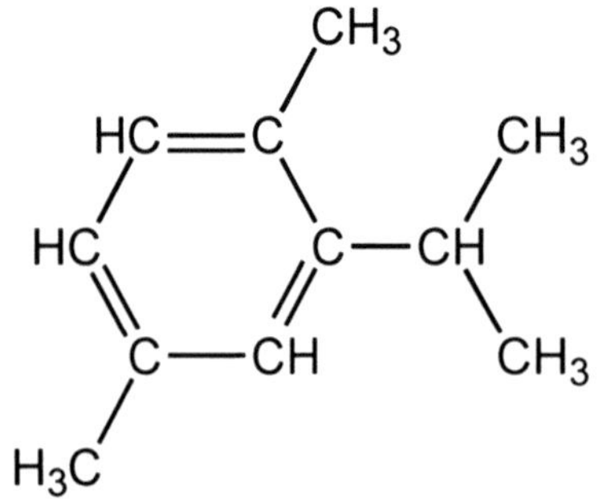

is 2-isopropyl-1,4-dimethylbenzene.

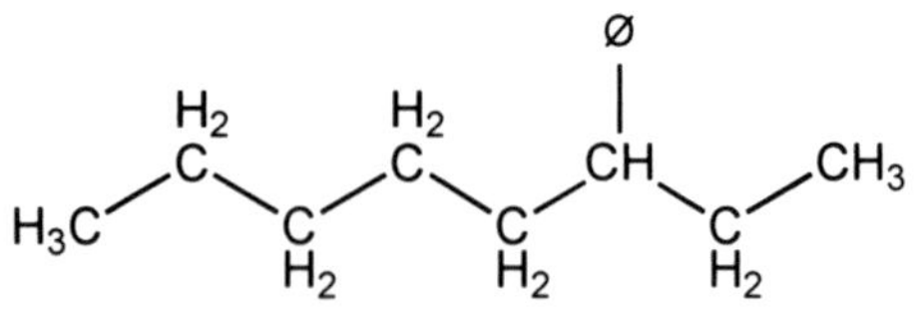

is 3-phenyloctane or (1-ethylhexyl)benzene.

The most often used common names for aromatic hydrocarbons are listed in the following table. Naphthalene is the simplest molecule formed by fused benzene rings.

Structure	Systematic name	Common name
$HC=CH$, HC (ring) $C-CH_3$, $HC=CH$	methylbenzene	toluene
$HC=CH$, HC (ring) $C-CH_3$, $HC=C$, CH_3	1,2-dimethylbenzene	ortho-xylene or o-xylene
$HC=CH$, HC (ring) $C-CH_3$, $C=CH$, H_3C	1,3-dimethylbenzene	meta-xylene or m-xylene
$HC=CH$, H_3C-C (ring) $C-CH_3$, $HC=CH$	1,4-dimethylbenzene	para-xylene or p-xylene
$HC=CH$, HC (ring) $C-CH$, CH_2, $HC=CH$	ethenylbenzene	styrene
fused rings HC, HC, CH, CH		naphthalene

Hydrocarbons consist entirely of nonpolar C-H bonds with no unpaired electrons. These compounds are relatively unreactive. The substitution of one or more atoms with unpaired electrons into the hydrocarbon backbone creates a **hydrocarbon derivative**. The unpaired electrons result in polar or charged portions of these molecules. These atoms fall into categories known as **functional groups**, and they create **local regions of reactivity**. Alkyl, alkenyl, alkynyl, and aryl groups may also be considered functional groups in some circumstances as described in the previous skill.

Two types of names are often used for the same hydrocarbon derivative. For **substitutive names**, the hydrocarbon name is written out and the correct prefix or suffix is added for the derivative group. For **functional class names**, the hydrocarbon is written as a functional group and the derivative is added as a second word.

Examples: CH_3Br may be called bromomethane (substitutive name) or methyl bromide (functional class name). CH_3CH_2OH is ethanol (substitutive name) or ethyl alcohol (functional class name).

In the tables on the following pages, the symbols R and R' represent hydrocarbons in covalent linkage to the functional group. Many derivatives are named in a similar manner to alkenes and alkynes, but the location and suffix of the functional group is used in place of –ene and –yne.

The **acyl group** is

There are non-systematic number roots for hydrocarbon derivatives containing acyl groups or derived from acyl groups. These are the ketones, carboxylic acids, esters, nitriles, and amides:

Number of carbons (including the acyl carbon)	Systematic prefix	Accepted prefix for acyl and nitrile functional groups
1	meth–	form–
2	eth–	acet–
3	prop–	propion–
4	but–	butyr–
5	pent–	pent–
6 and above	same for larger numbers	

Several functional groups utilize oxygen.

Class of molecule	Functional group	Structure	Affix	Exar
Alcohol	Hydroxyl ——OH	primary secondary tertiary	–ol	 ethanol
Ether	Oxy 		–oxy–	 methoxyethane or ethyl methyl ether
Aldehyde	Carbonyl 		–al	 propionaldehyde or propanal
Ketone			–one	 acetone
Carboxylic acid	Carboxyl 		–oic acid	 formic acid or methanoic acid

Class of molecule	Functional group	Structure	Affix	Example
Ester	Oxycarbonyl		–yl –oate	methyl butyrate or methyl butanoate
Acid anhydride	Carbonyloxycarbonyl		–oic anhydride	acetic anhydride or ethanoic anhydride

Derivatives utilizing other atoms are also common.

Class of molecule	Functional group	Structure		Affix	Example
Nitrile	Cyanide —C≡N	R—C≡N:		–nitrile	H_3C—C≡N acetonitrile or ethanonitrile
Amine	Amino —N—	primary: R—N(H)—H, H below	secondary: R_1—N—H, R_2 below; tertiary: R_1—N—R_3, R_2 below	–amine	H_3C—C(H_2)—NH_2 ethanamine
Amide	Aminocarbonyl	primary: O=C(R)—N(H)(H)	secondary: O=C(R_2)—N(R_1)(H); tertiary: O=C(R_3)—N(R_2)(R_1)	–amide	H_3C—C(H_2)—C(=O)—NH_2 propionamide or propanamide
Alkyl halide	Halide —X (where X is F, Cl, Br, or I)	R—X:		fluoro– chloro– bromo– iodo–	Br on CH, H_3C—CH—C(H_2)—CH_3 2-bromobutane

Hydrocarbon derivatives and functional groups may be identified using the tables found in the previous skill. One way to memorize the derivatives utilizing oxygen is to divide them into pairs with a hydrogen atom in the first element of the pair replaced by a hydrocarbon in the second element. These pairs are:

1) alcohol/ether.
2) aldehyde/ketone.
3) carboxylic acid/ester.

A final word about IUPAC nomenclature of organic compounds:

"IUPAC nomenclature" has become a synonym for "correct nomenclature," but the most recent IUPAC recommendations for the nomenclature of organic compounds are ignored in most textbooks and curricula in the United States. This situation is similar to the nomenclature for inorganic chemistry in that the older nomenclature is "systematic enough" to be taught and to avoid most errors, but is a poor choice to name new compounds in an unambiguous way.

Two important differences between the 1979 and more recent nomenclature involve the use of locant numbers:

1) The 1979 nomenclature usually places locant numbers before the name of the entire compound, but the 1993 recommendations suggest placing locant numbers immediately before the group that they represent.

2) The 1979 nomenclature always assigns a locant number of 1 to the attachment point of groups, but the 1993 nomenclature recommends stating the attachment point explicitly before "–yl–." This decreases the complexity of hydrocarbon residues.

COMPETENCY 0016 **UNDERSTAND FACTORS THAT AFFECT REACTION RATES AND METHODS OF MEASURING REACTION RATES.**

In order for one species to be converted to another during a chemical reaction, the reactants must collide. The collisions between the reactants determine how fast the reaction takes place. However, during a chemical reaction, only a fraction of the collisions between the appropriate reactant molecules convert them into product molecules. This occurs for two reasons:

1) Not all collisions occur with a **sufficiently high energy** for the reaction to occur.
2) Not all collisions **orient the molecules properly** for the reaction to occur.

The **activation energy**, E_a, of a reaction is the **minimum energy to overcome the barrier to the formation of products**. This is the minimum energy needed for the reaction to occur.

At the scale of individual molecules, a reaction typically involves a very small period of time when old bonds are broken and new bonds are formed. During this time, the molecules involved are in a **transition state** between reactants and products. A threshold of maximum energy is crossed when the arrangement of molecules is in an unfavorable intermediate between reactants and products known as the **activated complex**. Formulas and diagrams of activated complexes are often written within brackets to indicate they are transition states that are present for extremely small periods of time.

The activation energy, E_a, is the difference between the energy of reactants and the energy of the activated complex. The energy change during the reaction, ΔE, is the difference between the energy of the products and the energy of the reactants. The activation energy of the reverse reaction is $E_a - \Delta E$. These energy levels are represented in an **energy diagram** such as the one shown on the following page for $NO_2 + CO \rightleftharpoons NO + CO_2$.

This is an exothermic reaction because products are lower in energy than reactants.

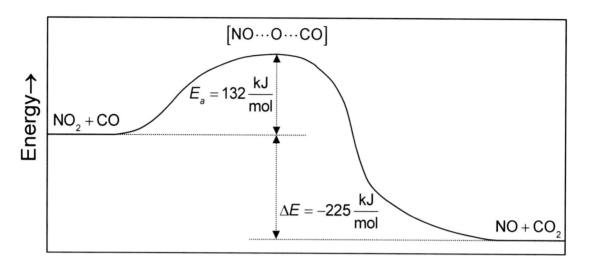

Reaction pathway→

An energy diagram is a conceptual tool, so there is some variability in how its axes are labeled. The y-axis of the diagram is usually labeled energy (E), but it is sometimes labeled "enthalpy (H)" or (rarely) "free energy (G)." There is an even greater variability in how the x-axis is labeled. The terms "reaction pathway," "reaction coordinate," "course of reaction," or "reaction progress" may be used on the x-axis, or the x-axis may remain without a label.

The energy diagrams of an endothermic and exothermic reaction are compared below.

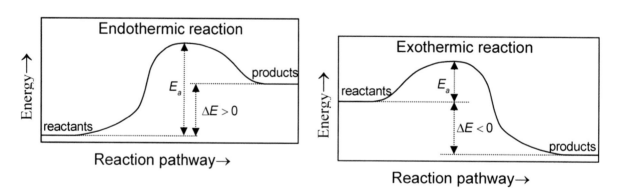

The rate of most simple reactions **increases with temperature** because a **greater fraction of molecules have the kinetic energy** required to overcome the reaction's activation energy. The chart below shows the effect of temperature on the distribution of kinetic energies in a sample of molecules. These curves are called **Maxwell-Boltzmann distributions**. The shaded areas represent the fraction of molecules containing sufficient kinetic energy for a reaction to occur. This area is larger at a higher temperature; so more molecules are above the activation energy and more molecules react per second.

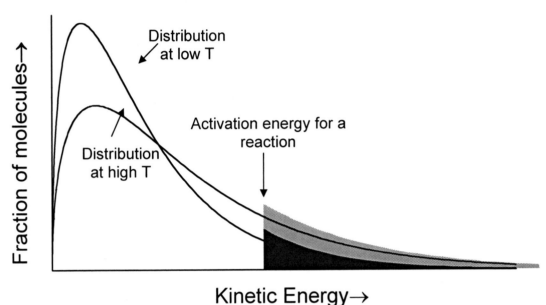

http://www.mhhe.com/physsci/chemistry/essentialchemistry/flash/activa2.swf provides an animated audio tutorial on energy diagrams.

Kinetic molecular theory may be applied to reaction rates in addition to physical constants like pressure. **Reaction rates increase with reactant concentration** because more reactant molecules are present and more are likely to collide with one another in a certain volume at higher concentrations. The nature of these relationships determines the rate law for the reaction. For ideal gases, the concentration of a reactant is its molar density, and this varies with pressure and temperature as discussed.

Kinetic molecular theory also predicts that **reaction rate constants (values for *k*) increase with temperature** because of two reasons:
1) More reactant molecules will collide with each other per second.
2) Each of these collisions will occur at a higher energy that is more likely to overcome the activation energy of the reaction.

A **catalyst** is a compound that increases the rate of a chemical reaction without changing itself permanently in the process.

Catalysts provide an alternate reaction mechanism for the reaction to proceed in the forward and in the reverse direction. Therefore, **catalysts have no impact on the chemical equilibrium** of a reaction. They will not make a less favorable reaction more favorable.

Catalysts reduce the activation energy of a reaction. This is the amount of energy needed for the reaction to begin. Molecules with such low energies that they would have taken a long time to react will react more rapidly if a catalyst is present.

The impact of a catalyst may also be represented on an energy diagram. **A catalyst increases the rate of both the forward and reverse reactions by lowering the activation energy** for the reaction. Catalysts provide a different activated complex for the reaction at a lower energy state.

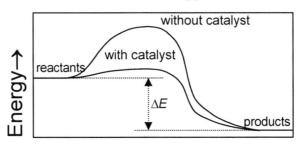

Reaction pathway$\rightarrow$

There are two types of catalysts: **Homogeneous catalysts** are in the same physical phase as the reactants. Biological catalysts are called **enzymes**, and most of them are homogeneous catalysts. A typical homogenous catalytic reaction mechanism involves an initial reaction with one reactant followed by a reaction with a second reactant and release of the catalyst:

$$A + C \rightarrow AC$$
$$B + AC \rightarrow AB + C$$
$$\text{Net reaction: } A + B \xrightarrow{\text{catalyst C}} AB$$

Heterogeneous catalysts are present in a different physical state from the reactants. A typical heterogeneous catalytic reaction involves a solid surface onto which molecules in a fluid phase temporarily attach themselves in such a way to favor a rapid reaction. Catalytic converters in cars utilize heterogeneous catalysis to break down harmful chemicals in exhaust.

Obtaining reaction rates from concentration data

The rate of any process is measured by its change per unit time. The speed of a car is measured by its change in position with time using units of kilometers/miles per hour.

The speed of a chemical reaction is usually measured by a change in the concentration of a reactant or product with time using units of **molarity per second** (M/s). The molarity of a chemical is represented in mathematical equations using brackets.

The **average reaction rate** is the change in concentration either reactant or product per unit time during a time interval:

$$\text{Average reaction rate} = \frac{\text{Change in concentration}}{\text{Change in time}}$$

Reaction rates are positive quantities. Product concentrations increase and reactant concentrations decrease with time, so a different formula is required depending on the identity of the component of interest:

$$\text{Average reaction rate} = \frac{\left[\text{product}\right]_{\text{final}} - \left[\text{product}\right]_{\text{initial}}}{\text{time}_{\text{final}} - \text{time}_{\text{iniial}}}$$

$$= \frac{\left[\text{reactant}\right]_{\text{initial}} - \left[\text{reactant}\right]_{\text{final}}}{\text{time}_{\text{final}} - \text{time}_{\text{iniial}}}$$

The **reaction rate** at any given time refers to the **instantaneous reaction rate**. This is found from the absolute value of the **slope of a curve of concentration vs. time**. An estimate of the reaction rate at time t may be found from the average reaction rate over a small time interval surrounding t. For those familiar with calculus notation, the following equations define reaction rate, but calculus is not needed for this skill:

$$\text{Reaction rate at time } t = \frac{d\left[\text{product}\right]}{dt} = -\frac{d\left[\text{reactant}\right]}{dt}.$$

Example: The following concentration data describes the decomposition of N_2O_5 according to the reaction $2N_2O_5 \rightarrow 4NO_2 + O_2$:

Time (sec)	[N$_2$O$_5$] (M)
0	0.0200
1000	0.0120
2000	0.0074
3000	0.0046
4000	0.0029
5000	0.0018
7500	0.0006
10000	0.0002

Determine the average reaction rate from 1000 to 5000 seconds and the instantaneous reaction rate at 0 and at 4000 seconds.

Solution: Average reaction rate from 0 to 7500 seconds is found from:

$$\frac{\left[\text{reactant}\right]_{\text{initial}} - \left[\text{reactant}\right]_{\text{final}}}{\text{time}_{\text{final}} - \text{time}_{\text{iniial}}} = \frac{0.0120\ M - 0.0018\ M}{5000\ \text{sec} - 1000\ \text{sec}} = 2.55 \times 10^{-6}\ \frac{M}{s}.$$

Instantaneous reaction rates are found by drawing lines tangent to the curve, finding the slopes of these lines, and forcing these slopes to be positive values.

At 0 seconds:

$$\text{rate=slope} = \frac{0.0200 \text{ M}}{2000 \text{ s}}$$

$$= 1.00 \times 10^{-5} \frac{\text{M}}{\text{s}}.$$

At 4000 seconds:

$$\text{rate=slope} = \frac{0.0090 \text{ M}}{6000 \text{ s}}$$

$$= 1.5 \times 10^{-6} \frac{\text{M}}{\text{s}}.$$

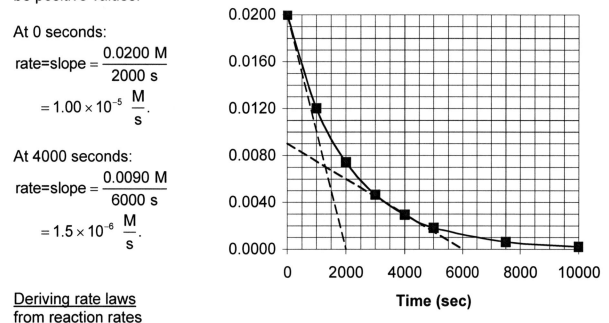

Time (sec)

Deriving rate laws
from reaction rates
A **rate law** is an **equation relating a reaction rate to concentration**. The rate laws for most reactions discussed in high-school level chemistry are of the form:

$$\text{Rate} = k[\text{reactant 1}]^{a}[\text{reactant 2}]^{b}\dots$$

In the above general equation, k is called the **rate constant**. a and b are called **reaction orders**. Most reactions considered in introductory chemistry have a reaction order of zero, one, or two. The sum of all reaction orders for a reaction is called the **overall reaction order**. Rate laws cannot be predicted from the stoichiometry of a reaction. They must be determined by experiment or derived from knowledge of reaction mechanism.

If a reaction is zero order for a reactant, the concentration of that reactant has no impact on rate as long as some reactant is present. If a reaction is first order for a reactant, the reaction rate is proportional to the reactant's concentration.

For a reaction that is second order with respect to a reactant, doubling that reactant's concentration increases reaction rate by a factor of four. Rate laws are determined by finding the appropriate reaction order describing **the impact of reactant concentration on reaction rate**.

Reaction rates typically have units of M/s (moles/liter-sec) and concentrations have units of M (moles/liter). For units to cancel properly in the expression above, the units found on the rate constant k must vary with overall reaction order as shown in the following table. The value of k may be determined by finding the slope of a plot charting a function of concentration against time. These functions may be memorized or computed using calculus.

Overall reaction order	Units of rate constant k	Method to determine k for rate laws with one reactant
0	M/sec	−(slope) of a chart of [reactant] vs. t
1	sec^{-1}	−(slope) of a chart of ln[reactant] vs. t
2	$M^{-1}sec^{-1}$	slope of a chart of 1/[reactant] vs. t

As an alternative to using the rate constant k, the course of **first order reactions** may be expressed in terms of a **half-life**, $t_{halflife}$. The half-life of a reaction is the time required for reactant concentration to reach half of its initial value. First order rate constants and half-lives are inversely proportional:

$$t_{halflife} = \frac{\ln 2}{k_{first\ order}} = \frac{0.693}{k_{first\ order}}$$

Example: Derive a rate law for the reaction $2N_2O_5 \rightarrow 4NO_2 + O_2$ using data from the previous example.

Solution: Three methods will be used to solve this problem.

1) In the previous example, we found the following two **instantaneous reaction rates**:

Time (sec)	[N$_2$O$_5$] (M)	Reaction rate (M/sec)
0	0.0200	1.00×10^{-5}
4000	0.0029	1.5×10^{-6}

A decrease in reactant concentration to 0.0029/0.0200=14.5% of its initial value led to a nearly proportional decrease in reaction rate to 15% of its initial value.

In other words, reaction rate remains proportional to reactant concentration. The reaction is first order:

$$\text{Rate} = k\left[N_2O_5\right].$$

We may estimate a value for the rate constant by dividing reaction rates by the concentration:

$$k_{first\ order} = \frac{\text{Rate}}{\left[N_2O_5\right]}.$$

Time (sec)	$[N_2O_5]$ (M)	Reaction rate (M/sec)	k (sec^{-1})
0	0.0200	1.00×10^{-5}	5.00×10^{-4}
4000	0.0029	1.5×10^{-6}	5.2×10^{-4}

2) We could estimate this rate constant by finding **average reaction rates** in each small time interval and assuming this rate occurs halfway between the two concentrations:

Time (sec)	$[N_2O_5]$ (M)	Average rate (M/sec)	Halfway $[N_2O_5]$ (M)	k (sec^{-1})
0	0.0200	8.00×10^{-6}	0.0160	5.00×10^{-4}
1000	0.0120	4.6×10^{-6}	0.0097	4.7×10^{-4}
2000	0.0074	2.8×10^{-6}	0.0060	4.7×10^{-4}
3000	0.0046	1.7×10^{-6}	0.0038	4.5×10^{-4}
4000	0.0029	1.1×10^{-6}	0.0024	4.7×10^{-4}
5000	0.0018	4.8×10^{-7}	0.0012	4.0×10^{-4}
7500	0.0006	2×10^{-7}	0.0004	4×10^{-4}
10000	0.0002			

3) If **concentration data** are given then no rate data needs to be found to determine a rate constant. For a first order reaction, chart the natural logarithm of concentration against time and find the slope.

Time (sec)	$[N_2O_5]$ (M)	$\ln[N_2O_5]$
0	0.0200	-3.91
1000	0.0120	-4.41
2000	0.0074	-4.90
3000	0.0046	-5.37
4000	0.0029	-5.83
5000	0.0018	-6.30
7500	0.0006	-7.39
10000	0.0002	-8.46

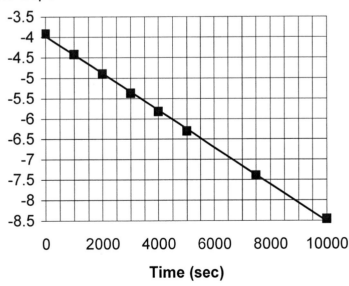

The slope may be determined from a best-fit method or it may be estimated from $\dfrac{-8.46-(-3.91)}{10000} = -5 \times 10^{-4}$.

The rate law describing this reaction is:

$$\text{Rate} = \left(5 \times 10^{-4} \, \frac{M}{\text{sec}}\right)[N_2O_5]$$

Derive rate laws from simple reaction mechanisms
A **reaction mechanism** is a series of **elementary reactions** that explain how a reaction occurs. These elementary reactions are also called elementary processes or elementary steps. **A reaction mechanism cannot be determined from reaction stoichiometry**. Stoichiometry indicates the number of molecules of reactants and products in an **overall reaction**. Elementary steps represent a **single event**. This might be a collision between two molecules or a single rearrangement of electrons within a molecule.

The simplest reaction mechanisms consist of a single elementary reaction. The number of molecules required determines the rate laws for these processes. For a **unimolecular process**:

$$A \rightarrow \text{products}$$

the number of molecules of A that decompose in a given time will be proportional to the number of molecules of A present. Therefore unimolecular processes are first order:

$$\text{Rate} = k[A].$$

For **bimolecular processes**, the rate law will be second order.

$$\text{For} \quad A + A \rightarrow \text{products}, \quad \text{Rate} = k[A]^2$$
$$\text{For} \quad A + B \rightarrow \text{products}, \quad \text{Rate} = k[A][B].$$

Most reaction mechanisms are multi-step processes involving **reaction intermediates**. Intermediates are chemicals that are formed during one elementary step and consumed during another, but they are not overall reactants or products. In many cases one elementary reaction in particular is the slowest and determines the overall reaction rate. This slowest reaction in the series is called the **rate-limiting step** or rate determining step.

Example: The reaction $NO_2(g) + CO(g) \rightarrow NO(g) + CO_2(g)$ is composed of the following elementary reactions in the gas phase:

$$NO_2 + NO_2 \rightarrow NO + NO_3$$
$$NO_3 + CO \rightarrow NO_2 + CO_2.$$

The first elementary reaction is very slow compared to the second. Determine the rate law for the overall reaction if NO_2 and CO are both present in sufficient quantity for the reaction to occur. Also name all reaction intermediates.

Solution: The first step will be rate limiting because it is slower. In other words, almost as soon as NO_3 is available, it reacts with CO, so the rate-limiting step is the formation of NO_3. The first step is bimolecular.

Therefore, the rate law for the entire reaction is: Rate=k$[NO_2]^2$.

NO_3 is formed during the first step and consumed during the second. NO_3 is the only reaction intermediate because it is neither a reactant nor a product of the overall reaction.

COMPETENCY 0017   UNDERSTAND THE PRINCIPLES OF CHEMICAL EQUILIBRIUM.

A dynamic equilibrium consists of two **opposing reversible processes** that both occur at the **same rate**. *Balance* is a synonym for equilibrium. A system at equilibrium is stable; it does not change with time. Equilibria are represented with double arrows.

When a process at equilibrium is observed, it often doesn't seem like anything is happening, but **at a microscopic scale, two events are taking place that balance each other**. An example is presented on the right. Arrows in this diagram represent the movement of molecules. When water is placed in a closed container, the water evaporates until the air in the container is saturated. After this occurs, the water level no longer changes, so an observer at the macroscopic scale would say that evaporation has stopped, but the reality on a microscopic scale is that both evaporation and condensation are taking place at the same rate. All equilibria between different phases of matter have this dynamic character on a microscopic scale.

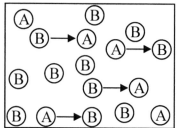

$$\text{Liquid} \underset{\text{condensation}}{\overset{\text{evaporation}}{\rightleftharpoons}} \text{Vapor}$$

Chemical reactions often do not "go to completion." Instead, products are generated from reactants up to a certain point when the reaction no longer seems to occur, leaving some reactant unaltered. At this point, the system is in a state of **chemical equilibrium** because **the rate of the forward reaction is equal to the rate of the reverse reaction**. An example is shown to the right. Arrows in this diagram represent the chemical reactions of individual molecules.

$$A \rightleftharpoons B$$

An observer at the macroscopic scale might say that no reaction is taking place at equilibrium, but at a microscopic scale, both the forward and reverse reactions are occurring at the same rate.

Homogeneous equilibrium refers to a chemical equilibrium among reactants and products that are all in the same phase of matter. **Heterogeneous equilibrium** takes place between two or more chemicals in different phases.

If a reaction at equilibrium is disturbed, changes occur to reestablish equilibrium. The nature of these changes is determined by **Le Chatelier's principle**.

A reaction at equilibrium contains a constant ratio of chemical species. The nature of these ratios is determined by an **equilibrium constant**.

The mathematical relationship between the concentrations of the reactants and products of a system is called the law of mass action. It governs equilibrium expressions.

Consider the general balanced reaction:

$$mA + nB \rightleftharpoons pR + qS$$

where m, n, p, and q are stoichiometric coefficients and A, B, R, and S are chemical species. An **equilibrium expression** relating to the concentrations of chemical species at equilibrium is determined by the equation:

$$K_{eq} = \frac{[R]^p [S]^q}{[A]^m [B]^n}$$

where K_{eq} is a constant value called the **equilibrium constant**. Product concentrations raised to the power of their stoichiometric coefficients are placed in the numerator and reactant concentrations raised to the power of their coefficients are placed in the denominator. Every reaction has a unique value of K_{eq} that varies only with temperature. Alternate subscripts are often given to the equilibrium constant. K_c or K with no subscript is often used instead of K_{eq} to represent the equilibrium constant. Other subscripts are used for specific reactions as described in the following skill.

Example: Write the equilibrium expression for the reaction
$2HI(g) \rightleftharpoons H_2(g) + I_2(g)$.

Solution: $K_{eq} = \dfrac{[H_2][I_2]}{[HI]^2}$.

The units associated with equilibrium constants in the expression above are molarity raised to the power of an integer that depends on the stoichiometric coefficients of the reaction, but it is common practice to write these constants as dimensionless values. Multiplying or dividing the equilibrium expression by 1 M as needed achieves these dimensionless values. **The equilibrium expression for a reaction written in one direction is the reciprocal of the expression for the reaction in the reverse direction**.

For a heterogeneous equilibrium (a chemical equilibrium with components in different phases), reactants or products may be pure liquids or solids. The concentration of a pure liquid or solid in moles/liter cannot change. It is a constant property of the material, and these constants are incorporated into the equilibrium constant. Therefore the concentrations of pure liquids and solids are absent from equilibrium expressions for heterogeneous equilibria.

Example: Write the equilibrium expression for the redox reaction between copper and silver: $Cu(s) + 2Ag^+(aq) \rightleftharpoons Cu^{2+}(aq) + 2Ag(s)$.

Solution: $K_{eq} = \dfrac{[Cu^{2+}]}{[Ag^+]^2}$. The solids do not appear in the equilibrium expression.

Calculation of unknown concentrations and concentration units

Many types of K_{eq} problems require the calculation of an unknown concentration at equilibrium from known quantities. These problems only require algebra to solve. Remember that equilibrium constants are forced to be dimensionless values. If all concentrations are represented in the same units, it is OK to be a little less cautious with units than for other problems.

Example: The reaction $N_2(g) + 3H_2(g) \rightleftharpoons 2NH_3(g)$ achieves equilibrium in the presence of 0.27 M H_2 and 0.094 M N_2. What is the concentration of ammonium under these conditions if the reaction at the given temperature has an equilibrium constant of $K_{eq} = 0.11$?

Solution: First we will solve this problem without worrying about units because all concentrations are given in M:

$$K_{eq} = \frac{[NH_3]^2}{[N_2][H_2]^3} = \frac{[NH_3]^2}{(0.094)(0.27)^3} = 0.11.$$

Solving for [NH₃] yields: $[NH_3] = \sqrt{(0.11)(0.094)(0.27)^3} = 0.014$ M.

This is the preferred method for solving these problems.

If units are to be treated rigorously then a more explicit definition of K_{eq} is written. This assures us that we achieve a dimensionless value for K_{eq} by repeatedly multiplying by 1 M:

$$K_{eq} = \frac{[NH_3]^2(1\,M)^2}{[N_2][H_2]^3} = \frac{[NH_3]^2(1\,M)^2}{(0.094\,M)(0.27\,M)^3} = 0.11.$$

Solving for [NH₃] yields: $[NH_3] = \sqrt{(0.11)\dfrac{(0.094\,M)(0.27\,M)^3}{(1\,M)^2}} = 0.014$ M.

Determining the direction of a reaction

Some K_{eq} problems give every concentration value for a reaction that is not at equilibrium and ask which direction the reaction will proceed for a given equilibrium constant. Solving these problems is a two step process:

1) Insert the non-equilibrium reaction concentrations into the equilibrium expression to obtain a **reaction quotient**, Q.
2) If $Q<K_{eq}$, there are too many reactant molecules for the products, and the reaction proceeds to the right. If $Q>K_{eq}$, there are too many product molecules for the reactants, and the reaction proceeds to the left. If $Q=K_{eq}$, the reaction is at equilibrium.

Example: Predict the direction in which the reaction $H_2(g)+I_2(g)\rightleftharpoons 2HI(g)$ will proceed if initial concentrations are 0.004 mM H_2, 0.006 mM I_2, and 0.011 mM HI given $K_{eq}=48$.

Solution: 1) The reactant quotient is $Q=\dfrac{[HI]^2}{[H_2][I_2]}=\dfrac{(0.011)^2}{(0.004)(0.006)}=5$.

2) The reactant quotient is less than K_{eq}. That is, 5<48. Therefore, the numerator (products) of the reaction quotient is too small for the denominator (reactants). The trend towards equilibrium with time will increase the numerator relative to the denominator until a ratio of 48 is achieved. More reactants will turn into products and the reaction will proceed to the right.

A look at the Q/K ratio will give an indication of the status of the system:

Q/K:	
> 1	Product concentration too high for equilibrium; net reaction proceeds to left.
= 1	System is at equilibrium; no net change will occur.
< 1	Product concentration too low for equilibrium; net reaction proceeds to right.

Special equilibrium constants

A few equilibrium constants are used often enough to have their own unique nomenclature.

The **solubility-product constant**, K_{sp}, is the equilibrium constant for an ionic solid in contact with a saturated aqueous solution. The two processes with equal rates in this case are dissolution and crystallization.

$$\text{ionic compound}(s) \rightleftharpoons p \text{ Cation}^+(aq) + q \text{ Anion}^-(aq)$$

$$K_{sp} = \left[\text{cation}^+\right]^p \left[\text{anion}^-\right]^q.$$

This is an example of a heterogeneous equilibrium, so the concentration of pure solid is not included as a variable. K_{sp} is a different quantity from solubility. K_{sp} is an equilibrium constant, and solubility is the mass of solid that is able to dissolve in a given quantity of water.

Example: Solid lead chloride $PbCl_2$ is allowed to dissolve in pure water until equilibrium has been reached and the solution is saturated. The concentration of Pb^{2+} is 0.016 M. What is K_{sp} for $PbCl_2$?

Solution: For $PbCl_2(s) \rightleftharpoons Pb^{2+}(aq) + 2Cl^-(aq)$, $K_{sp} = \left[Pb^{2+}\right]\left[Cl^-\right]^2$. The only source of both ions in solution is $PbCl_2$, so the concentration of Cl^- must be twice that for Pb^{2+}, or 0.032 M. Therefore,

$$K_{sp} = (0.016)(0.032)^2 = 1.6 \times 10^{-5}.$$

The **acid-dissociation constant**, K_a, is the equilibrium constant for the ionization of a weak acid to a hydrogen ion and its conjugate base:

$$HX(aq) \rightleftharpoons H^+(aq) + X^-(aq) \qquad K_a = \frac{\left[H^+\right]\left[X^-\right]}{[HX]}.$$

Polyprotic acids have unique values for each dissociation: K_{a1}, K_{a2}, etc.

Example: Hydrofluoric acid is dissolved in pure water until $[H^+]$ reaches 0.006 M. What is the concentration of undissociated HF? K_a for HF is 6.8×10^{-4}.

Solution: For $HF(aq) \rightleftharpoons H^+(aq) + F^-(aq)$, $K_a = \frac{\left[H^+\right]\left[F^-\right]}{[HF]}$. The principle source of both ions is dissociation of HF (autoionization of water is negligible). Therefore $\left[F^-\right] = \left[H^+\right] = 0.006$ M, and

$$[HF] = \frac{\left[H^+\right]\left[F^-\right]}{K_a} = \frac{(0.006)^2}{6.8 \times 10^{-4}} = 0.05 \text{ M}.$$

The **base-dissociation constant**, K_b, is the equilibrium constant for the addition of a proton to a weak base by water to form its conjugate acid and an OH^- ion. In these reactions, it is water that is dissociating as a result of reaction with the base:

$$\text{weak base}(aq) + H_2O(l) \rightleftharpoons \text{conjugate acid}(aq) + OH^-(aq)$$

$$K_b = \frac{[\text{conjugate acid}][OH^-]}{[\text{weak base}]}.$$

The concentration of water is nearly constant and is incorporated into the dissociation constant.

For ammonia (the most common weak base), the equilibrium reaction and base-dissociation constant are:

$$NH_3(aq) + H_2O(l) \rightleftharpoons NH_4^+(aq) + OH^-(aq)$$

$$K_b = \frac{[NH_4^+][OH^-]}{[NH_3]}$$

Example: K_b for ammonia at 25 °C is 1.8×10^{-5}. What is the concentration of OH^- in an ammonia solution at equilibrium containing 0.2 M NH_3 at 25°C?

Solution: Call $x = [OH^-]$. The principle source of both ions is NH_3 (autoionization of water is negligible). Therefore $x = [NH_3^+]$ also.

$$K_b = \frac{[NH_4^+][OH^-]}{[NH_3]} = \frac{x^2}{0.2} = 1.8 \times 10^{-5}.$$

Solving for x yields: $x = \sqrt{(0.2)(1.8 \times 10^{-5})} = 0.002$ M OH^-

The **ion-product constant for water**, K_w is the equilibrium constant for the dissociation of H_2O. Water molecules may donate protons to other water molecules in a process known as autoionization:

$$2H_2O(l) \rightleftharpoons H_3O^+(aq) + OH^-(aq)$$

A hydrated water molecule is often referred to as H^+ (aq), so the above equation may be rewritten as the following reaction that defines K_w. As with K_b, the concentration of water is nearly constant.

$$H_2O(l) \rightleftharpoons H^+(aq) + OH^-(aq)$$
$$K_w = [H^+][OH^-] = 1.0 \times 10^{-14} \text{ at } 25°C.$$

Example: a) What is the concentration of OH^- in an aqueous solution with an H^+ concentration of 2.5×10^{-6} M?
   b) What is the concentration of H^+ when pure water reaches equilibrium?

Solution: a) $K_w = [H^+][OH^-] = (2.5 \times 10^{-6})[OH^-] = 1.0 \times 10^{-14}$.

   Solving for $[OH^-]$ yields $[OH^-] = \dfrac{1.0 \times 10^{-14}}{2.5 \times 10^{-6}} = 4.0 \times 10^{-9}$.

   b) The autoionization of pure water creates an equal concentration of the two ions:

   $[H^+] = [OH^-]$. Therefore $K_w = [H^+]^2 = 1.0 \times 10^{-14}$.

   Solving for $[H^+]$ yields $[H^+] = 1.0 \times 10^{-7}$.

A system at equilibrium is in a state of balance because forward and reverse processes are taking place at equal rates. If equilibrium is disturbed by changing concentration, pressure, or temperature, the state of balance is affected for a period of time before the equilibrium shifts to achieve a new state of balance. **Le Chatelier's principle states that equilibrium will shift to partially offset the impact of an altered condition**.

Change in reactant and product concentrations
If a chemical reaction is at equilibrium, Le Chatelier's principle predicts that **adding a substance**—either a reactant or a product—will shift the reaction so **a new equilibrium is established by consuming some of the added substance**. Removing a substance will cause the reaction to move in the direction that forms more of that substance.

Example: The reaction $CO + 2H_2 \rightleftharpoons CH_3OH$ is used to synthesize methanol. Equilibrium is established, and then additional CO is added to the reaction vessel. Predict the impact on each reaction component after CO is added.

Solution: Le Chatelier's principle states that the reaction will shift to partially offset the impact of the added CO. Therefore, CO concentration will decrease, and the reaction will "shift to the right." H_2 concentration will also decrease and CH_3OH concentration will increase.

Change in pressure for gases

If a chemical reaction is at equilibrium in the gas phase, Le Chatelier's principle predicts that **an increase in pressure** will shift the reaction so **a new equilibrium is established by decreasing the number of gas moles present**. A decrease in the number of moles partially offsets this rise in pressure. Decreasing pressure will cause the reaction to move in the direction that forms more moles of gas. These changes in pressure might result from altering the volume of the reaction vessel at constant temperature.

Example: The reaction $N_2 + 3H_2 \rightleftharpoons 2NH_3$ is used to synthesize ammonia. Equilibrium is established. Next the reaction vessel is expanded at constant temperature. Predict the impact on each reaction component after this expansion occurs.

Solution: The expansion will result in a decrease in pressure. Le Chatelier's principle states that the reaction will shift to partially offset this decrease by increasing the number of moles present. There are 4 moles on the left side of the equation and 2 moles on the right, so the reaction will shift to the left. N_2 and H_2 concentration will increase. NH_3 concentration will decrease.

Change in temperature

Le Chatelier's principle predicts that **when heat is added** at constant pressure to a system at equilibrium, **the reaction will shift in the direction that absorbs heat** until a new equilibrium is established. For an endothermic process, the reaction will shift to the right towards product formation. For an exothermic process, the reaction will shift to the left towards reactant formation. If you understand the application of Le Chatelier's principle to concentration changes then writing "heat" on the appropriate side of the equation will help you understand its application to changes in temperature.

Example: $N_2 + 3H_2 \rightleftharpoons 2NH_3$ is an exothermic reaction. First equilibrium is established and then the temperature is decreased. Predict the impact of the lower temperature on each reaction component.

Solution: Since the reaction is exothermic, we may write it as: $N_2 + 3H_2 \rightleftharpoons 2NH_3 + Heat$. For the purpose of finding the impact of temperature on equilibrium processes, we may consider heat as if it were a reaction component. Le Chatelier's principle states that after a temperature decrease, the reaction will shift to partially offset the impact of a loss of heat. Therefore more heat will be produced, and the reaction will shift to the right. N_2 and H_2 concentration will decrease. NH_3 concentration will increase.

A flash animation with audio that demonstrates Le Chatelier's principle is at http://www.mhhe.com/physsci/chemistry/essentialchemistry/flash/lechv17.swf.

Example: The following questions refer to a slightly exothermic reversible reaction where NO_2 is a brown gas and N_2O_4 is colorless. Select the correct response.

$$2\ NO_2\ (g) \rightleftharpoons N_2O_4\ (g)$$

a. An increase in temperature would make the solution

    (A) brown   (B) colorless

    (C) another color   (D) no changes in the color

b. An increase in pressure would make the solution

    (A) brown   (B) colorless

    (C) another color   (D) no changes in the color

c. Additional NO_2 would make the solution

    (A) brown   (B) colorless

    (C) another color   (D) no changes in the color

d. Removing N_2O_4 would make the solution

    (A) brown   (B) colorless

    (C) another color   (D) no changes in the color

e. A decrease in the concentration of NO_2 would make the solution

(A) brown   (B) colorless

(C) another color   (D) no changes in the color

Solution:

a. An increase in temperature would make the solution

(A) brown   (B) colorless

(C) another color   (D) no changes in the color

Rewrite the equation showing energy on the product side since it is slightly exothermic.

$$2\ NO_2\ (g) \leftrightarrows N_2O_4\ (g) + E$$

Now increasing energy would make the system want to use up the extra energy in order to re-establish equilibrium so the concentration of N_2O_4 would decrease and the concentration of NO_2 would increase. This would make the reaction appear brown.

b. An increase in pressure would make the solution

(A) brown   (B) colorless

(C) another color   (D) no changes in the color

Solution:

Referring to the equation: $2\ NO_2\ (g) \leftrightarrows N_2O_4\ (g)$
There are fewer moles of molecules on the product side so to relieve the stress of increasing pressure, the system would produce more product and have fewer molecules present. This would make the reaction appear less brown or colorless.

c. Additional NO_2 would make the solution

(A) brown

(B) colorless

(C) another color

(D) no changes in the color

Solution: Adding more N_2O_4 to the reaction would stress the system at equilibrium and cause it to make more NO_2 using up the excess N_2O_4 in the process. Since more NO_2 would be present the reaction would appear brown.

d. Removing N_2O_4 would make the solution

(A) brown

(B) colorless

(C) another color

(D) no changes in the color

Solution: Removing N_2O_4 from a system at equilibrium would stress the system. In order to re-establish equilibrium, the concentration of N_2O_4 would need to increase to replace the N_2O_4 that was removed. The system would work to make more N_2O_4, decreasing the concentration of NO_2 and the system would become lighter brown to colorless.

COMPETENCY 0018 UNDERSTAND THE THEORIES, PRINCIPLES, AND APPLICATIONS OF ACID-BASE CHEMISTY.

It was recognized centuries ago that many substances could be divided into the two general categories. **Acids** have a sour taste (as in lemon juice), dissolve many metals, and turn litmus paper red. **Bases** have a bitter taste (as in soaps), feel slippery, and turn litmus paper blue. The chemical reaction between an acid and a base is called **neutralization**. The products of neutralization reactions are neither acids nor bases. Litmus paper is an example of an **acid-base indicator**, a substance that changes color when added to an acid to a base.

Properties of Acids and Bases	
Acids	Bases
Produce H+ (as H_3O+) ions in water (the hydronium ion is a hydrogen ion attached to a water molecule)	Produce OH- ions in water
Taste sour	Taste bitter, chalky
Corrode metals	Are electrolytes
Electrolytes	Feel soapy, slippery
React with bases to form a salt and water	React with acids to form salts and water
pH is less than 7	pH greater than 7
Turns blue litmus paper to red "Blue to Red A-CID"	Turns red litmus paper to blue "Basic Blue"
React with metals to produce hydrogen gas	React with fats to produce soap (saponification)

Arrhenius definition of acids and bases

Svante **Arrhenius** proposed in the 1880s that **acids form H^+ ions and bases form OH^- ions in water**. The net ionic reaction for neutralization between an Arrhenius acid and base always produces water as shown below for nitric acid and sodium hydroxide:

$$HNO_3(aq) + NaOH(aq) \rightarrow NaNO_3(aq) + H_2O(l)$$

$$H^+(aq) + NO_3^-(aq) + Na^+(aq) + OH^-(aq) \rightarrow NO_3^-(aq) + Na^+(aq) + H_2O(l) \text{ (complete ionic)}$$

$$H^+(aq) + OH^-(aq) \rightarrow H_2O(l) \text{ (net ionic)}$$

The H$^+$(aq) ion

In acid-base systems, **"protonated water"** or **"H$^+$(aq)"** are shorthand for a **mixture of water ions**. For example, HCl reacting in water may be represented as a dissociation:

$$HCl(aq) \rightarrow H^+(aq) + Cl^-(aq)$$

The same reaction may be described as the transfer of a proton to water to form H$_3$O$^+$:

$$HCl(aq) + H_2O(l) \rightarrow Cl^-(aq) + H_3O^+(aq)$$

H$_3$O$^+$ is called a **hydronium ion**. Its Lewis structure is shown below to the left. In reality, the hydrogen bonds in water are so strong that H$^+$ ions exist in water as a mixture of species in a hydrogen bond network. Two of them are shown below at center and to the right. Hydrogen bonds are shown as dashed lines.

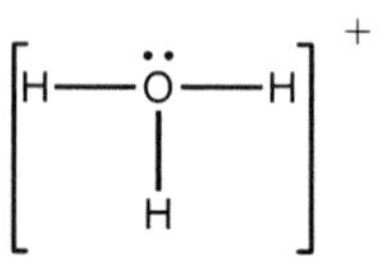

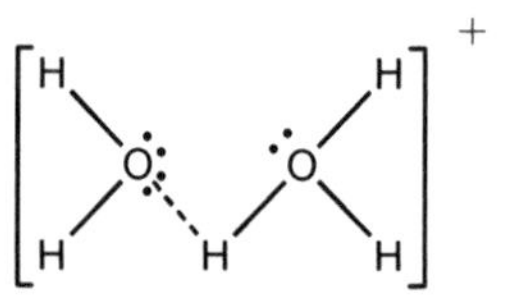

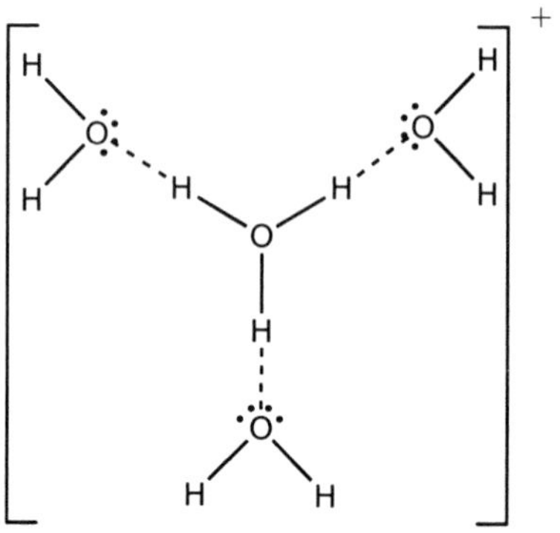

Brønsted-Lowry definition of acids and bases

In the 1920s, Johannes **Brønsted** and Thomas **Lowry** recognized that **acids can transfer a proton to bases** regardless of whether an OH$^-$ ion accepts the proton. In an equilibrium reaction, the direction of proton transfer depends on whether the reaction is read left to right or right to left, so **Brønsted acids and bases exist in conjugate pairs with and without a proton**. Acids that are able to transfer more than one proton are called **polyprotic acids**.

Examples:
1) In the reaction:

$$HF(aq) + H_2O(l) \rightleftharpoons F^-(aq) + H_3O^+(aq),$$

HF transfers a proton to water. Therefore HF is the Brønsted acid and H$_2$O is the Brønsted base. But in the reverse direction, hydronium ions transfer a proton to fluoride ions. H$_3$O$^+$ is the conjugate acid of H$_2$O because it has an additional proton, and F$^-$ is the conjugate base of HF because it lacks a proton.

2) In the reaction:

$$NH_3(aq) + H_2O(l) \rightleftharpoons NH_4^+(aq) + OH^-(aq),$$

water transfers a proton to ammonia. H_2O is the Brønsted acid and OH^- is its conjugate base. NH_3 is the Brønsted base and NH_4^+ is its conjugate acid.

3) In the reaction:

$$H_3PO_4 + HS^- \rightleftharpoons H_2PO_4^- + H_2S$$

$H_3PO_4/H_2PO_4^-$ is one conjugate acid-base pair and H_2S/HS^- is the other.

4) H_3PO_4 is a polyprotic acid. It may further dissociate to transfer more than one proton:

$$H_3PO_4 \rightleftharpoons H_2PO_4^- + H^+$$
$$H_2PO_4^- \rightleftharpoons HPO_4^{2-} + H^+$$
$$HPO_4^{2-} \rightleftharpoons PO_4^{3-} + H^+$$

Example: In the following reactions, identify the acid/conjugate base pair and the base conjugate acid pair.
a) $HNO_3 + OH^- \rightarrow$
b) $CH_3NH_2 + H_2O \rightarrow$
c) $OH^- + HPO_4^{-2} \rightarrow$

Solution: a) $HNO_3 + OH^- \rightarrow H_2O + NO_3^-$

HNO_3 and NO_3^- make one pair (acid-conjugate base). The hydrogen from the HNO_3 is being donated to the OH^- ion to make H_2O, leaving NO_3^-. Therefore, the HNO_3 is acting as an acid and the OH^- is acting as the base. Once the hydrogen has been added to the OH^- ion, it now has the ability to donate a proton (H^+) to the NO_3^- ion making water a proton donor (or Bronsted-Lowry acid) and the NO_3^- a base (proton acceptor).

OH^- and H_2O make the other (base-conjugate acid)

b) $CH_3NH_2 + H_2O \rightarrow CH_3NH_3^+ + OH^-$

CH_3NH_2 and $CH_3NH_3^+$ make one pair (base-conjugate acid pair). In this reaction, H_2O is donating a proton making it an acid. CH_3NH_2 is accepting the proton to become $CH_3NH_3^+$, making it a base. On the product side, the $CH_3NH_3^+$ has the ability to donate a proton to the OH^-, making $CH_3NH_3^+$ an acid and OH^- a base.

OH^- and H_2O make the other (conjugate base-acid pair)

c) $OH^- + HPO_4^{-2} \rightarrow H_2O + PO_4^{-3}$

HPO_4^{-2} and PO_4^{-3} make one pair. Here, HPO_4^{2-} is acting as a Bronsted-Lowry base when it donates a proton to OH^-. When the OH^- accepts the proton to become H_2O, it is acting as a base. PO_3^{3-} is the conjugate base in that it can accept a proton to form HPO_4^{2-}. The proton would come from the H_2O molecule on the product side, making water the conjugate acid.

OH^- and H_2O make the other.

Strong and weak acids and bases

Strong acids and bases are strong electrolytes, and weak acids and bases are weak electrolytes, so **strong acids and bases completely dissociate in water**, but weak acids and bases do not.

Example: $HCl(aq) + H_2O(l) \rightarrow H_3O^+(aq) + Cl^-(aq)$ goes to completion because HCl is a strong acid. The acids in the examples on the previous page were all weak.

The aqueous dissociation constants K_a and K_b quantify acid and base strength. Another way of looking at acid dissociation is that strong acids transfer protons more readily than H_3O^+ transfers protons, so they protonate water, the conjugate base of H_3O^+. In general, **if two acid/base conjugate pairs are present, the stronger acid will transfer a proton to the conjugate base of the weaker acid**.

Acid and base **strength is not related to safety**. Weak acids like HF may be extremely corrosive and dangerous.

The most **common strong acids and bases** are listed in the following table:

Strong acid		Strong base	
HCl	Hydrochloric acid	LiOH	Lithium hydroxide
HBr	Hydrobromic acid	NaOH	Sodium hydroxide
HI	Hydroiodic acid	KOH	Potassium hydroxide
HNO_3	Nitric acid	$Ca(OH)_2$	Calcium hydroxide
H_2SO_4	Sulfuric acid	$Sr(OH)_2$	Strontium hydroxide
$HClO_4$	Perchloric acid	$Ba(OH)_2$	Barium hydroxide

A flash animation tutorial demonstrating the difference between strong and weak acids is located at
http://www.mhhe.com/physsci/chemistry/essentialchemistry/flash/acid13.swf.

Trends in acid and base strength

The strongest acid in a polyprotic series is always **the acid with the most protons** (e.g. H_2SO_4 is a stronger acid than HSO_4^-). The strongest acid in a series with the same central atom is always **the acid with the central atom at the highest oxidation number** (e.g. $HClO_4 > HClO_3 > HClO_2 > HClO$ in terms of acid strength). The strongest acid in a series with different central atoms at the same oxidation number is usually **the acid with the central atom at the highest electronegativity** (e.g. the K_a of $HClO > HBrO > HIO$).

Lewis definition of acids and bases

The transfer of a proton from a Brønsted acid to a Brønsted base requires that the base accept the proton. When Lewis diagrams are used to draw the proton donation of Brønsted acid-base reactions, it is always clear that the base must contain an unshared electron pair to form a bond with the proton. For example, ammonia contains an unshared electron pair in the following reaction:

$$
H^+ \; + \; :\!\!\overset{\displaystyle H}{\underset{\displaystyle H}{N}}\!\!-\!H \; \longrightarrow \; \left[H\!-\!\overset{\displaystyle H}{\underset{\displaystyle H}{N}}\!\!-\!H \right]^+
$$

In the 1920s, Gilbert N. **Lewis** proposed that **bases donate unshared electron pairs to acids**, regardless of whether the donation is made to a proton or to another atom. Boron trifluoride is an example of a Lewis acid that is not a Brønsted acid because it is a chemical that accepts an electron pair without involving an H^+ ion:

$$
\overset{\displaystyle F}{\underset{\displaystyle F}{B}}\!\!-\!F \; + \; :\!\!\overset{\displaystyle H}{\underset{\displaystyle H}{N}}\!\!-\!H \; \longrightarrow \; F\!-\!\overset{\displaystyle F}{\underset{\displaystyle F}{B}}\!\!-\!\overset{\displaystyle H}{\underset{\displaystyle H}{N}}\!\!-\!H
$$

The Lewis theory of acids and bases is more general than Brønsted-Lowry theory, but Brønsted-Lowry's definition is used more frequently. The terms "acid" and "base" most often refer to Brønsted acids and bases, and the term "Lewis acid" is usually reserved for chemicals like BF_3 that are not Brønsted acids.

Summary of definitions

A Lewis base transfers an electron pair to a Lewis acid. A Brønsted acid transfers a proton to a Brønsted base. These exist in conjugate pairs at equilibrium. In an Arrhenius base, the proton acceptor (electron pair donor) is OH^-. All Arrhenius acids/bases are Brønsted acids/bases and all Brønsted acids/bases are Lewis acids/bases. Each definition contains a subset of the one that comes after it.

The concentration of $H^+(aq)$ ions is often expressed in terms of pH. **The pH of a solution is the negative base-10 logarithm of the hydrogen-ion molarity.**

$$pH = -\log[H^+] = \log\left(\frac{1}{[H^+]}\right).$$

A ten-fold increase in $[H^+]$ decreases the pH by one unit. $[H^+]$ may be found from pH using the expression:

$$[H^+] = 10^{-pH}.$$

The concentration of H^+ ions has been shown as $[H^+] = 10^{-7}$ M for pure water with $[H^+] = [OH^-]$. Thus **the pH of a neutral solution is 7**. In an **acidic solution**, $[H^+] > 10^{-7}$ M and **pH < 7**. In a basic solution, $[H^+] < 10^{-7}$ M and **pH > 7**.

The negative base-10 log is a convenient way of representing other small numbers used in chemistry by placing the letter "p" before the symbol. Values of K_a are often represented as pK_a, with $pK_a = -\log K_a$. The concentration of OH^- (aq) ions may also be expressed in terms of pOH, with $pOH = -\log[OH^-]$.

The ion-product constant of water, $K_w = [H^+][OH^-] = 1.0 \times 10^{-14}$ at 25 °C. The value of K_w can used to determine the relationship between pH and pOH by taking the negative log of the expression:

$$-\log K_w = -\log[H^+] - \log[OH^-] = -\log(10^{-14}).$$

$$\text{Therefore: } pH + pOH = 14.$$

Example: An aqueous solution has an H^+ ion concentration of 4.0×10^{-9}. Is the solution acidic or basic? What is the pH of the solution? What is the pOH?

Solution: The solution is basic because $[H^+] < 10^{-7}$ M.

$$pH = -\log[H^+] = -\log 4 \times 10^{-9}$$

$$= 8.4.$$

$$pH + pOH = 14. \text{ Therefore } pOH = 14 - pH = 14 - 8.4 = 5.6.$$

A **buffer solution** is a solution that **resists a change in pH** after addition of small amounts of an acid or a base. Buffer solutions require the presence of an acid to neutralize an added base and also the presence of a base to neutralize an added acid. These two components present in the buffer also must not neutralize each other! A **conjugate acid-base pair is present in buffers** to fulfill these requirements.

Buffers are prepared by mixing together **a weak acid or base and a salt of the acid or base** that provides the conjugate.

Consider the buffer solution prepared by mixing together acetic acid— $HC_2H_3O_2$ —and sodium acetate— $C_2H_3O_2^-$ containing Na^+ as a spectator ion. The equilibrium reaction for this acid/conjugate base pair is:

$$HC_2H_3O_2 \rightleftharpoons C_2H_3O_2^- + H^+.$$

If H^+ ions from a strong acid are added to this buffer solution, Le Chatelier's principle predicts that the reaction will shift to the left and much of this H^+ will be consumed to create more $HC_2H_3O_2$ from $C_2H_3O_2^-$. If a strong base that consumes H^+ is added to this buffer solution, Le Chatelier's principle predicts that the reaction will shift to the right and much of the consumed H^+ will be replaced by the dissociation of $HC_2H_3O_2$. The net effect is that **buffer solutions prevent large changes in pH that occur when an acid or base is added to pure water** or to an unbuffered solution.

The amount of acid or base that a buffer solution can neutralize before dramatic pH changes begins to occur is called its **buffering capacity**. Blood and seawater both contain several conjugate acid-base pairs to buffer the solution's pH and decrease the impact of acids and bases on living things.

An excellent flash animation with audio to explain the action of buffering solutions is found at
http://www.mhhe.com/physsci/chemistry/essentialchemistry/flash/buffer12.swf.

Standard titration

In a typical acid-base **titration, an acid-base indicator** (such as *phenolphthalein*) or a **pH meter** is used to monitor the course of a **neutralization reaction**. The usual goal of titration is to **determine an unknown concentration** of an acid (or base) by neutralizing it with a known concentration of base (or acid).

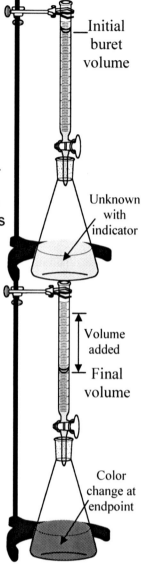

Initial buret volume

The reagent of known concentration is usually used as the **titrant**. The titrant is poured into a **buret** (also spelled *burette*) until it is nearly full, and an initial buret reading is taken. Buret numbering is close to zero when nearly full. A known volume of the solution of unknown concentration is added to a flask and placed under the buret. The indicator is added or the pH meter probe is inserted. The initial state of a titration experiment is shown to the right above.

The buret's stopcock is opened and titrant is slowly added until the solution permanently changes color or the pH rapidly changes. This is the **end point of titration**, and a final buret reading is made. The final state of a titration experiment is shown to the right below.

Unknown with indicator

The endpoint occurs when the number of **acid and base equivalents in the flask are identical**:

$$N_{acid} = N_{base}. \text{ Therefore, } C_{acid}V_{acid} = C_{base}V_{base}.$$

The endpoint is also known as the titration **equivalence point**.

Volume added

Final volume

Titration data typically consist of:

$$V_{inital} \Rightarrow \text{Initial buret volume} \quad V_{final} \Rightarrow \text{Final buret volume}$$

$$C_{known} \Rightarrow \text{Concentration of known solution}$$

$$V_{unknown} \Rightarrow \text{Volume of unknown solution.}$$

Color change at endpoint

To determine the unknown concentration, first find the volume of titrant at the known concentration added: $V_{known} = V_{final} - V_{initial}$.

At the equivalence point, $N_{unknown} = N_{known}$.

$$\text{Therefore, } C_{unknown} = \frac{C_{known}V_{known}}{V_{unknown}} = \frac{C_{known}\left(V_{final} - V_{initial}\right)}{V_{unknown}}.$$

Units of molarity may be used for concentration in the above expressions **unless a mole of either solution yields more than one acid or base equivalent.** In that case, concentration must be expressed using **normality.**

Example: A 20.0 mL sample of an HCl solution is titrated with 0.200 M NaOH. The initial buret reading is 1.8 mL and the final buret reading at the titration endpoint is 29.1 mL. What is the molarity of the HCl sample?

Solution: Two solution methods will be used. The first method is better for those who are good at unit manipulations and less skilled at memorizing formulas. HCl contains one acid equivalent and NaOH contains one base equivalent, so we may use molarity in all our calculations.

1) Calculate the moles of the known substance added to the flask:

$$0.200 \frac{mol}{L} \times \frac{1\,L}{1000\,mL} \times \left(29.1\,mL - 1.8\,mL\right) = 0.00546 \text{ mol NaOH.}$$

At the endpoint, this base will neutralize 0.00546 mol HCl. Therefore, this amount of HCl must have been present in the sample before the titration.

$$\frac{0.00546 \text{ mol HCl}}{0.0200\,L} = 0.273 \text{ M HCl.}$$

2) Utilize the formula: $C_{unknown} = \dfrac{C_{known}\left(V_{final} - V_{initial}\right)}{V_{unknown}}.$

$$C_{HCl} = \frac{C_{NaOH}\left(V_{final} - V_{initial}\right)}{V_{HCl}} = \frac{0.200 \text{ M}\left(29.1\,mL - 1.8\,mL\right)}{20.0\,mL} = 0.273 \text{ M HCl.}$$

Titrating with the unknown
In a common variation of standard titration, the unknown is added to the buret as a titrant and the reagent of known concentration is placed in the flask. The chemistry involved is the same as in the standard case, and the mathematics is also identical except for the identity of the two volumes. For this variation, V_{known} will be the volume added to the flask before titration begins and $V_{unknown} = V_{final} - V_{initial}$. Therefore:

$$C_{unknown} = \frac{C_{known}V_{known}}{V_{unknown}} = \frac{C_{known}V_{known}}{V_{final} - V_{initial}}.$$

Example: 30.0 mL of a 0.150 M HNO_3 solution is titrated with $Ca(OH)_2$. The initial buret volume is 0.6 mL and the final buret volume at the equivalent point is 22.2 mL. What is the molarity of $Ca(OH)_2$ used for the titration?

Solution: The same two solution methods will be used as in the previous example. 1 mol $Ca(OH)_2$ contains 2 base equivalents because it reacts with 2 moles of H^+ via the reaction

$Ca(OH)_2 + 2HNO_3 \rightarrow Ca(NO_3)_2 + 2H_2O$. Therefore, normality must be used in the formula for solution method 2.

1) First, calculate the moles of the substance in the flask:

$$0.150 \frac{mol}{L} \times 0.0300 \text{ L} = 0.00450 \text{ mol } HNO_3.$$

This acid must be titrated with 0.00450 mol base equivalents for neutralization to occur at the end point. We calculate moles $Ca(OH)_2$ used in the titration from stoichiometry:

$$0.00450 \text{ base equivalents} \times \frac{1 \text{ mol } Ca(OH)_2}{2 \text{ base equivalents}} = 0.00225 \text{ mol } Ca(OH)_2.$$

The molarity of $Ca(OH)_2$ is found from the volume used in the titration:

$$\frac{0.00225 \text{ mol } Ca(OH)_3}{0.0222 \text{ L} - 0.0006 \text{ L}} = 0.104 \text{ M } Ca(OH)_3.$$

2) Utilize the formula: $C_{unknown} = \frac{C_{known} V_{known}}{V_{final} - V_{initial}}$ using units of normality.

For HNO_3, molarity=normality because 1 mol contains 1 acid equivalent.

$$C_{Ca(OH)_2} = \frac{C_{HNO_3} V_{HNO_3}}{V_{final} - V_{initial}} = \frac{\left(0.150 \text{ M} \times \frac{1 \text{ N}}{1 \text{ M}}\right)(30.0 \text{ mL})}{22.2 \text{ mL} - 0.6 \text{ mL}} = 0.208 \text{ N } Ca(OH)_2.$$

This value is converted to molarity. For $Ca(OH)_2$, normality is twice molarity because 1 mol contains 2 base equivalents.

$$0.208 \text{ N } Ca(OH)_2 \times \frac{1 \text{ M } Ca(OH)_2}{2 \text{ N } Ca(OH)_2} = 0.104 \text{ M } Ca(OH)_2.$$

Interpreting titration curves

A **titration curve** is a plot of a solution's **pH charted against the volume of an added acid or base**. Titration curves are obtained if a pH meter is used to monitor the titration instead of an indicator. At the equivalence point, the titration curve is nearly vertical. This is the point where a rapid change in pH occurs. In addition to determining the equivalence point, the **shape of titration curves** may be interpreted to determine **acid/base strength and the presence of a polyprotic acid.**

The pH at the equivalence point of a titration is the **pH of the salt solution obtained when the amount of acid is equal to amount of base**. For a strong acid and a strong base, the equivalence point occurs at the neutral pH of 7. For example, an equimolar solution of HCl and NaOH will contain NaCl(*aq*) at its equivalence point.

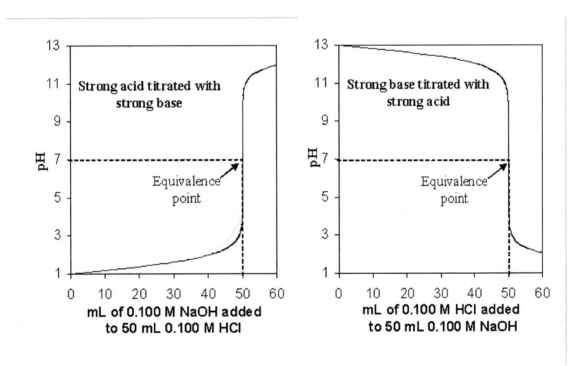

The salt solution at **the equivalence point of a titration involving a weak acid or base will not be at neutral pH**. For example, an equimolar solution of NaOH and hypochlorous acid HClO at the equivalence point of a titration will be a base because it is indistinguishable from a solution of sodium hypochlorite. A pure solution of NaClO(*aq*) will be a base because the ClO⁻ ion is the conjugate base of HClO, and it consumes $H^+(aq)$ in the reaction: $ClO^- + H^+ \rightleftharpoons HClO$.

In a similar fashion, an equimolar solution of HCl and NH_3 will be an acid because a solution of $NH_4Cl(aq)$ is an acid. It generates $H^+(aq)$ in the reaction: $NH_4^+ \rightleftharpoons NH_3 + H^+$.

Contrast the following **titration curves for a weak acid or base** with those for a strong acid and strong base in the preceding page:

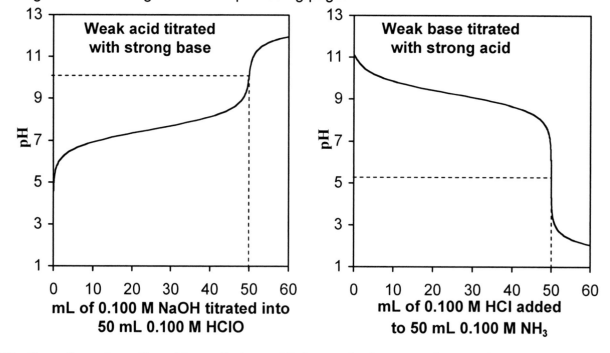

Titration of a polyprotic acid results in **multiple equivalence points** and a curve with more "bumps" as shown below for sulfurous acid and the carbonate ion.

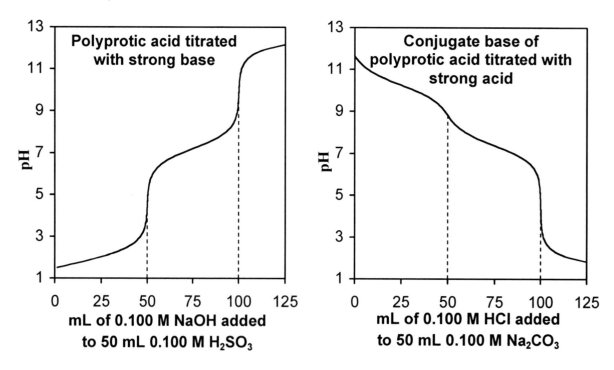

COMPETENCY 0019   UNDERSTAND REDOX REACTIONS AND ELECTROCHEMISTRY.

Redox is shorthand for *reduction* and *oxidation*. **Reduction** is the **gain of an electron** by a molecule, atom, or ion. **Oxidation** is the **loss of an electron** by a molecule, atom, or ion. These two processes always occur together. Electrons lost by one substance are gained by another. In a redox process, the **oxidation numbers** of atoms are altered. Reduction decreases the oxidation number of an atom. Oxidation increases the oxidation number.

The easiest redox processes to identify are those involving monatomic ions with altered charges. For example, the reaction

$$Zn(s) + Cu^{2+}(aq) \rightarrow Zn^{2+}(aq) + Cu(s)$$

is a redox process because electrons are transferred from Zn to Cu.

However, many redox reactions involve the transfer of electrons from one molecular compound to another. In these cases, **oxidation numbers must be determined** as follows:

Oxidation numbers, sometimes called oxidation states, are signed numbers assigned to atoms in molecules and ions. They allow us to keep track of the electrons associated with each atom. Oxidation numbers are frequently used to write chemical formulas, to help us predict properties of compounds, and to help balance equations in which electrons are transferred. Knowledge of the oxidation state of an atom gives us an idea about its positive or negative character. In themselves, oxidation numbers have no physical meaning; they are used to simplify tasks that are more difficult to accomplish without them.

The Rules:

1. **Free elements are assigned an oxidation state of 0.**

   e.g. Al, Na, Fe, H_2, O_2, N_2, Cl_2 etc have zero oxidation states.

2. **The oxidation state for any simple one-atom ion is equal to its charge.**

   e.g. the oxidation state of Na^+ is +1, Be^{2+}, +2, and of F^-, -1.

3. **The alkali metals (Li, Na, K, Rb, Cs and Fr) in compounds are always assigned an oxidation state of +1.**

   e.g. in LiOH (Li, +1), in Na_2SO_4(Na, +1).

4. **Fluorine in compounds is always assigned an oxidation state of -1.**

   e.g. in HF_2^-, BF_2^-.

5. **The alkaline earth metals (Be, Mg, Ca, Sr, Ba, and Ra) and also Zn and Cd in compounds are always assigned an oxidation state of +2. Similarly, Al and Ga are always +3.**

   e.g. in $CaSO_4$(Ca, +2), $AlCl_3$ (Al, +3).

6. **Hydrogen in compounds is assigned an oxidation state of +1. Exception - Hydrides, e.g. LiH (H=-1).**

   e.g. in H_2SO_4 (H, +1).

7. **Oxygen in compounds is assigned an oxidation state of -2. Exception - Peroxide, e.g. H_2O_2 (O = -1).**

   e.g. in H_3PO_4 (O, -2).

8. **The sum of the oxidation states of all the atoms in a species must be equal to net charge on the species.**

   e.g. Net Charge of $HClO_4$ = 0, i.e. [+1(H)+7(Cl)-2*4(O)] = 0

   Net Charge of CrO_4^{2-}=-2,

   To solve Cr's oxidation state: x - 4*2(O) = -2, x = +6, so the oxidation state of Cr is +6.

For example, the reaction

$$H_2 + F_2 \rightarrow 2HF$$

is a redox process because the oxidation numbers of atoms are altered. The oxidation numbers of elements are always zero, and oxidation numbers in a compound are never zero. Fluorine is the more electronegative element, so in HF it has an oxidation number of −1 and hydrogen has an oxidation number of +1. This is a redox process where electrons are transferred from H_2 to F_2 to create HF.

In the reaction

$$HCl + NaOH \rightarrow NaCl + H_2O,$$

the H-atoms on both sides of the reaction have an oxidation number of +1, the atom of Cl has an oxidation number of −1, the Na-atom has an oxidation number of +1, and the atom of O has an oxidation number of −2. **This is not a redox process because oxidation numbers remain unchanged** by the reaction.

An **oxidizing agent** (also called an oxidant or oxidizer) has the ability to oxidize other substances by removing electrons from them. The **oxidizing agent is reduced** in the process. A **reducing agent** (also called a reductive agent, reductant or reducer) is a substance that has the ability to reduce other substances by transferring electrons to them. The **reducing agent is oxidized** in the process.

Redox reactions may always be written as **two half-reactions**, a **reduction half-reaction** with **electrons as a reactant** and an **oxidation half-reaction** with **electrons as a product**.

For example, the redox reactions considered in the previous skill:

$$Zn(s) + Cu^{2+}(aq) \rightarrow Zn^{2+}(aq) + Cu(s) \quad \text{and} \quad H_2 + F_2 \rightarrow 2HF$$

may be written in terms of the half-reactions:

$$2e^- + Cu^{2+}(aq) \rightarrow Cu(s) \qquad 2e^- + F_2 \rightarrow 2F^-$$
$$\text{and}$$
$$Zn(s) \rightarrow Zn^{2+}(aq) + 2e^-. \qquad H_2 \rightarrow 2H^+ + 2e^-.$$

An additional (non-redox) reaction, $2F^- + 2H^+ \rightarrow 2HF$, achieves the final products for the second reaction.

Determining whether a chemical equation is balanced (see section 0023) requires an additional step for redox reactions because there must be a **balance of charge**. For example, the equation:

$$Sn^{2+} + Fe^{3+} \rightarrow Sn^{4+} + Fe^{2+}$$

contains one Sn and one Fe on each side but it is not balanced because the sum of charges on the left side of the equation is +5 and the sum on the right side is +6. One electron is gained in the reduction half-reaction ($Fe^{3+} + e^- \rightarrow Fe^{2+}$), but two are lost in the oxidation half-reaction ($Sn^{2+} \rightarrow Sn^{4+} + 2e^-$).

The equation:

$$Sn^{2+} + 2Fe^{3+} \rightarrow Sn^{4+} + 2Fe^{2+}$$

is properly balanced because both sides contain the same sum of charges (+8) and electrons cancel from the half-reactions:

$$2Fe^{3+} + 2e^- \rightarrow 2Fe^{2+}$$
$$Sn^{2+} \rightarrow Sn^{4+} + 2e^-.$$

Oxidation Number Method:

Redox reactions must be balanced to observe the Law of Conservation of Mass. This process is a little more complicated than balancing other reactions because the number of electrons lost must equal the number of electrons gained. Balancing redox reactions, then, conserves not only mass but also charge or electrons. It can be accomplished by slightly varying our balancing process.

$$Cr_2O_3(s) + Al(s) \longrightarrow Cr(s) + Al_2O_{3(s)}$$

Assign oxidation numbers to identify which atoms are losing and gaining electrons.

$$Cr_2O_3(s) + Al(s) \longrightarrow Cr(s) + Al_2O_{3(s)}$$
$$3+ \ 2- \qquad 0 \qquad 0 \qquad 3+ \ 2-$$

Identify those atoms gaining and losing electrons:

$$Cr^{3+} \longrightarrow Cr^0 \quad \text{gained 3 electrons : reduction}$$
$$Al^0 \longrightarrow Al^{3+} \quad \text{lost 3 electrons: oxidation}$$

Balance the atoms and electrons:

$$Cr_2O_3(s) \longrightarrow 2Cr(s) + 6 \text{ electrons}$$
$$2Al(s) + 6 \text{ electrons} \longrightarrow Al_2O_3(s)$$

Balance the half reactions by adding missing elements. Ignore elements whose oxidation number does not change. Add H_2O for oxygen and H^+ for hydrogen.

$$Cr_2O_3(s) \longrightarrow 2Cr(s) + 6 \text{ electrons } + \mathbf{3\ H_2O}$$

Need 3 oxygen atoms on product side. This requires $6H^+$ on the reactant side.

$$Cr_2O_3(s) + \mathbf{6\ H^+} \longrightarrow 2Cr(s) + 6 \text{ electrons } + \mathbf{3\ H_2O}$$

AND

$$2Al(s) + 6 \text{ electrons } ^+ \mathbf{3\ H_2O} \longrightarrow Al_2O_{3(s)} {}^+ \mathbf{6\ H^+}$$

Need 3 oxygen atoms on reactant side. This requires $6H^+$ on the product side.

Put the two half reactions together and add the species. Cancel out the species that occur in both the reactants and products.

$$Cr_2O_3(s) + 6\ H^+ {}^+ 2Al(s) + 6 \text{ electrons } ^+ 3\ H_2O \longrightarrow 2Cr(s) + 6 \text{ electrons } + 3\ H_2O + Al_2O_{3(s)} {}^+ 6\ H^+$$

The balanced equation is:

$$Cr_2O_3(s) + 2Al(s) \longrightarrow 2Cr(s) + Al_2O_{3(s)}$$

Try another: $AgNO_3 + Cu \longrightarrow CuNO_3 + Ag$

Assign oxidation numbers to identify which atoms are losing and gaining electrons.

$$AgNO_3 + Cu \longrightarrow Cu(NO_3)_2 + Ag$$
$$1+ \ 5+ \ 2- \ \ 0 \qquad 2+ \ 5+ \ 2- \ \ 0$$

Identify those atoms gaining and losing electrons:

$$Ag^{1+} + 1 \ e^- \longrightarrow \ Ag^0 \quad \text{1 electron gained: reduction}$$
$$Cu^0 \longrightarrow \ Cu^{2+} + 2 \ e^- \quad \text{2 electrons lost: oxidation}$$

Balance the atoms and electrons:

$$AgNO_3 + 1 \ e^- \longrightarrow \ Ag^0$$
$$Cu^0 \longrightarrow \ Cu(NO_3)_2 \ + 2 \ e^-$$
$$Cu^0 \longrightarrow \ Cu^{2+} + 2 \ e^-$$

Balance the electrons:

$$AgNO_3 + 1 \ e^- \longrightarrow \ Ag^0$$
$$Cu^0 \longrightarrow \ Cu(NO_3)_2 \ + 2 \ e^-$$

1 electron gained and 2 electrons lost. Needs to be equal so 2 electrons need to be gained.

$$2 \ [Ag + 1 \ e^- \longrightarrow \ Ag^0] = \textbf{2 AgNO}_3 + \textbf{2 e}^- \longrightarrow \textbf{2 Ag}^{0+}$$

Reduction: $2 \ AgNO_3 + 2 \ e^- \longrightarrow 2 \ Ag^{0+}$

Oxidation: $\quad Cu^0 \longrightarrow \ Cu(NO_3)_2 \ + 2 \ e^-$

Balance the half reactions by adding missing elements. Ignore elements whose oxidation number does not change. Add H_2O for oxygen and H^+ for hydrogen

$$2 \ AgNO_3 + 2 \ e^- \longrightarrow 2 \ Ag^{0+} \ \textbf{+ 2 NO}_3$$
$$\textbf{2NO}_3 + \ \ Cu^0 \longrightarrow \ Cu(NO_3)_2 \ + 2 \ e^-$$

Put the two half reactions together and add the species. Cancel out the species that occur in both the reactants and products.

Reduction: $2 \ AgNO_3 + 2 \ e^- \longrightarrow 2 \ Ag^{0+} \ + 2 \ NO_3$

Oxidation: $2NO_3 + \ \ Cu^0 \longrightarrow \ Cu(NO_3)_2 \ + 2 \ e^-$

The balanced reaction is:

$2 \ AgNO_3 + Cu \longrightarrow \ Cu(NO_3)_2 + 2 \ Ag$

Try this one: $Ag_2S + HNO_3 \longrightarrow AgNO_3 + NO + S + H_2O$

Assign oxidation numbers:

$$Ag_2S + HNO_3 \longrightarrow AgNO_3 + NO + S + H_2O$$

1+ 2- 1+ 5+ 2-   1+ 5+ 2-   2+ 2-   0   1+ 2-

Identify those atoms gaining and losing electrons:

oxidation: $S^{2-} \longrightarrow S^0 + 2\,e^-$

reduction: $N^{5+} + 3\,e^{-+} \longrightarrow N^{2+}$

Balance the atoms:

Oxidation: $2\,NO_3 + Ag_2S \longrightarrow S + 2\,e^- + 2\,AgNO_3$

Reduction: $3\,H^+ + HNO_3 + 3\,e^- \longrightarrow NO + 2\,H_2O$

Balance electrons lost and gained:

2 electrons lost and 3 electron gained. Need to be equal so find multiple: 6

Oxidation: $3[\,2\,NO_3 + Ag_2S \longrightarrow S + 2\,e^- + 2\,AgNO_3]$

$6\,NO_3 + 3\,Ag_2S \longrightarrow 3\,S + 6\,e^- + 6\,AgNO_3$

Reduction: $2[\,3\,H^+ + HNO_3 + 3\,e^- \longrightarrow NO + 2\,H_2O]$

$6\,H^+ + 2\,HNO_3 + 6\,e^- \longrightarrow 2\,NO + 4\,H_2O$

Put the two half reactions together and add the species. Cancel out the species that occur in both the reactants and products.

$6\,NO_3 + 3\,Ag_2S + 6\,H^+ + 2\,HNO_3 + 6\,e^- \longrightarrow 3\,S + 6\,e^- + 6\,AgNO_3 + 2\,NO + 4\,H_2O$

The balanced equation is:

$$3\,Ag_2S + 8\,HNO_3 \longrightarrow 6\,AgNO_3 + 2\,NO + 3\,S + 4\,H_2O$$

To balance a redox reaction which occurs in a basic solution is a very similar to balancing a redox reaction which occurs in acidic conditions. First, balance the reaction as you would for an acidic solution and then adjust for the basic solution. Here is an example using the half-reaction method:

Solid chromium(III) hydroxide, $Cr(OH)_3$, reacts with aqueous chlorate ions, ClO_3^-, in basic conditions to form chromate ions, CrO_4^{2-}, and chloride ions, Cl^-.

$$Cr(OH)_3(s) + ClO_3^-(aq) \rightarrow CrO_4^{2-}(aq) + Cl^-(aq) \quad (basic)$$

1. Write the half-reactions:

   $Cr(OH)_3(s) \rightarrow CrO_4^{2-}(aq)$ and

   $ClO_3^-(aq) \rightarrow Cl^-(aq)$

2. Balance the atoms in each half-reaction. Use H_2O to add oxygen atoms and H^+ to add hydrogen atoms.

   $H_2O(l) + Cr(OH)_3(s) \rightarrow CrO_4^{2-}(aq) + 5 H^+(aq)$

   $6H^+(aq) + ClO_3^-(aq) \rightarrow Cl^-(aq) + 3 H_2O(l)$

3. Balance the charges of both half-reactions by adding electrons.

   $H_2O(l) + Cr(OH)_3(s) \rightarrow CrO_4^{2-}(aq) + 5 H^+(aq)$

   has a charge of +3 on the right and 0 on the left. Adding 3 electrons to the right side will give that side a 0 charge as well.

   $H_2O(l) + Cr(OH)_3(s) \rightarrow CrO_4^{2-}(aq) + 5 H^+(aq) + 3e^-$

   $6H^+(aq) + ClO_3^-(aq) \rightarrow Cl^-(aq) + 3 H_2O(l)$

   has a charge of -1 on the right side and a +5 on the left. Six electrons need to be added to the left side to equal the -1 charge on the right side.

   $6 e^- + 6H^+(aq) + ClO_3^-(aq) \rightarrow Cl^-(aq) + 3 H_2O(l)$

4. The number of electrons lost must equal the number of electrons gained so multiply each half-reactions by a number that will give equal numbers of electrons lost and gained.

$$H_2O \text{ (l)} + Cr(OH)_3(s) \rightarrow CrO_4^{2-} \text{ (aq)} + 5 H^+ \text{ (aq)} + 3e^-$$

$$6 e^- + 6H^+ \text{ (aq)} + ClO_3^- \text{ (aq)} \rightarrow Cl^- \text{ (aq)} + 3 H_2O \text{ (l)}$$

The first half reaction needs to be multiplied by 2 to equal the 6 electrons gained in the second half-reaction.

$$2[H_2O \text{ (l)} + Cr(OH)_3(s) \rightarrow CrO_4^{2-} \text{ (aq)} + 5 H^+ \text{ (aq)} + 3e^-] =$$

$$2H_2O \text{ (l)} + 2 Cr(OH)_3(s) \rightarrow 2 CrO_4^{2-} \text{ (aq)} + 10 H^+ \text{ (aq)} + 6e^-$$

5. Add the two half-reactions together; canceling out species that appear on both sides of the reaction.

$$2H_2O + 2\!\!\!/ Cr(OH)_3(s) + 6\!\!\!/e^- + 6H\!\!\!/^+(aq) + ClO_3(aq) \rightarrow 2 CrO_4^{2-} \text{ (aq)} + 1\!\!\!/0 H^+ \text{ (aq)} + 6\!\!\!/e^- + Cl(aq) + 3\!\!\!/H_2O$$

$$4 H^+ \text{ (aq)} \qquad 1 H_2O \text{ (l)}$$

6. Since the reaction occurs in basic solution and there are 4 H^+ ions on the right side, 4 OH^- need to be added to both sides. Combine the H^+ and OH^- where appropriate to make water molecules.

$$4 OH^- \text{ (aq)} + 2 Cr(OH)_3(s) + ClO_3(aq) \rightarrow 2 Cr(aq) + 4 H^+ \text{ (aq)} + Cl(aq) + 1H_2O \text{ (l)} + 4 OH^- \text{ (aq)}$$

$$4 H_2O \text{ (l)}$$

7. Write final balanced equation:

$$4 OH^- \text{ (aq)} + 2 Cr(OH)_3(s) + ClO_3^- \text{(aq)} \rightarrow 2 CrO_4^{2-} \text{ (aq)} + Cl^- \text{ (aq)} + 5 H_2O \text{ (l)}$$

57 videos of redox experiments are presented here:
http://chemmovies.unl.edu/chemistry/redoxlp/redox000.html.

Electrolytic cells use electricity to force nonspontaneous redox reactions to occur. **Electrochemical cells generate electricity** by permitting spontaneous redox reactions to occur. The two types of cells have some components in common.

Both systems contain two **electrodes**. An electrode is a piece of conducting metal that is used to make contact with a nonmetallic material. One electrode is an **anode**. An **oxidation reaction occurs at the anode**, so electrons are removed from a substance there. The other electrode is a **cathode.** A **reduction reaction occurs at the cathode**, so electrons are added to a substance there. Electrons flow from anode to cathode outside either device.

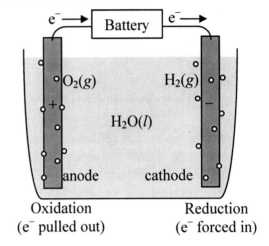

Electrolytic systems

Electrolysis is a chemical process **driven by a battery** or another source of electromotive force. This source pulls electrons out of the chemical process at the anode and forces electrons in the cathode. The result is a **negatively charged cathode and a positively charged anode**.

Oxidation (e⁻ pulled out) Reduction (e⁻ forced in)

Electrolysis of pure water forms O_2 bubbles at the anode by the oxidation half-reaction:

$$2H_2O(l) \rightarrow 4H^+(aq) + O_2(g) + 4e^-$$

and forms H_2 bubbles at the cathode by the reduction half-reaction:

$$2H_2O(l) + 2e^- \rightarrow H_2(g) + 2OH^-(aq).$$

The net redox reaction is:

$$2H_2O(l) \rightarrow 2H_2(g) + O_2(g).$$

Neither electrode took part in the reaction described above. An electrode that is only used to contact the reaction and deliver or remove electrons is called an **inert electrode**. An electrode that takes part in the reaction is called an **active electrode**.

Electroplating is the process of **depositing dissolved metal cations** in a smooth even coat onto an object used as an active electrode. Electroplating is used to protect metal surfaces or for decoration. For example, to electroplate a copper surface with nickel, a nickel rod is used for the anode and the copper object is used for the cathode. $NiCl_2(aq)$ or another substance with free nickel ions is used in the electrolytic cell. $Ni(s) \rightarrow Ni^{2+}(aq) + 2e^-$ occurs at the anode and $Ni^{2+}(aq) + 2e^- \xrightarrow{\text{onto Cu}} Ni(s)$ occurs at the cathode.

Electrochemical systems

An **electrochemical cell** separates the half-reactions of a redox process into two compartments or half-cells. Electrochemical cells are also called *galvanic cells* or *voltaic cells*.

A **battery** consists of one or more electrochemical cells connected together. Electron transfer from the oxidation to the reduction reaction may only take place through an external circuit.

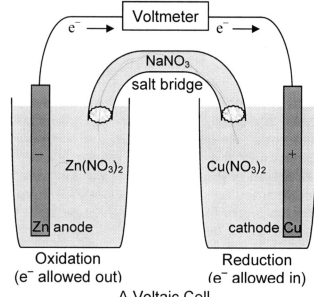

A Voltaic Cell

Electrochemical systems provide a **source of electromotive force**. This force is also called or *voltage* or *cell potential* and is measured in **volts**. Electrons are allowed to leave the chemical process at the anode and permitted to enter at the cathode. The result is a **negatively charged anode and a positively charged cathode**.

Electrical neutrality is maintained in the half-cells by **ions migrating** through a **salt bridge**. A salt bridge in the simplest cells is an inverted U-tube filled with a non-reacting electrolyte and plugged at both ends with a material like cotton or glass wool that permits ion migration but prevents the electrolyte from falling out.

The spontaneous redox reaction $Zn(s) + Cu^{2+}(aq) \rightarrow Zn^{2+}(aq) + Cu(s)$ generates a voltage in the cell above. The oxidation half-reaction $Zn(s) \rightarrow Zn^{2+}(aq) + 2e^-$ occurs at the anode. Electrons are allowed to flow through a voltmeter before they are consumed by the reduction half-reaction $Cu^{2+}(aq) + 2e^- \rightarrow Cu(s)$ at the cathode. Zinc dissolves away from the anode into solution, and copper from the solution builds up onto the cathode.

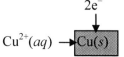

To maintain electrical neutrality in both compartments, positive ions (Zn^{2+} and Na^+) migrate through the salt bridge from the anode half-cell to the cathode half-cell and negative ions (NO_3^-) migrate in the opposite direction.

An animation of the cell described above is located at http://www.mhhe.com/physsci/chemistry/essentialchemistry/flash/galvan5.swf. A summary of anode and cathode properties for both cell types is contained in the table below.

		Electrolytic cell	Electrochemical cell
Anode	Half-reaction	Oxidation	Oxidation
	Electron flow	Pulled out	Allowed out
	Electrode polarity	+	−
Cathode	Half-reaction	Reduction	Reduction
	Electron flow	Forced in	Allowed in
	Electrode polarity	−	+

The reducing and oxidizing agents in a standard electrochemical cell are depleted with time. In a **rechargeable battery** (e.g., lead storage batteries in cars) the direction of the spontaneous redox reaction is reversed and **reactants are regenerated** when electrical energy is added into the system. A **fuel cell** has the same components as a standard electrochemical cell except that **reactants are continuously supplied**.

A **standard cell potential, E°_{cell}**, is the voltage generated by an electrochemical cell at **100 kPa and 25 °C** when all components of the reaction are pure materials or solutes at a **concentration of 1 M**. Older textbooks may use 1 atm instead of 100 kPa. Standard solute concentrations may differ from 1 M for solutions that behave in a non-ideal way, but this difference is beyond the scope of general high school chemistry.

Standard cell potentials are calculated from the **sum of the two half-reaction potentials** for the reduction and oxidation reactions occurring in the cell:

$$E^{\circ'}_{cell} = E^{\circ}_{red}(\text{cathode}) + E^{\circ}_{ox}(\text{anode}$$

All half-reaction potentials are relative to the reduction of H^+ to form H_2. This potential is assigned a value of zero:

$$\text{For } 2H^+(aq \text{ at } 1 \text{ M}) + 2e^- \rightarrow H_2(g \text{ at } 100 \text{ kPa}), \quad E^{\circ}_{red} = 0 \text{ V.}$$

The standard potential of an oxidation half-reaction E°_{ox} **is equal in magnitude but has the opposite sign to the potential of the reverse reduction reaction**. Standard half-cell potentials are **tabulated as reduction potentials**. These are sometimes referred to as **standard electrode potentials $E°$**. Therefore,

$$E^{\circ}_{cell} = E^{\circ}(\text{cathode}) - E^{\circ}(\text{anode}.$$

Example: Given $E°=0.34$ V for $Cu^{2+}(aq)+2e^-\rightarrow Cu(s)$ and
$E°= -0.76$ V for $Zn^{2+}(aq)+2e^-\rightarrow Zn(s)$, what is the standard cell
potential of the $Zn(s)+Cu^{2+}(aq) \rightarrow Zn^{2+}(aq)+Cu(s)$ system ?

Solution: $E°_{cell} = E°(\text{cathode}) - E°(\text{anode})$

$$= E°\left(Cu^{2+}(aq)+2e^- \rightarrow Cu(s)\right) - E°\left(Zn^{2+}(aq)+2e^- \rightarrow Zn(s)\right)$$

$$= 0.34 \text{ V} - (-0.76 \text{ V}) = 1.10 \text{ V}.$$

Spontaneity

When the value of $E°$ is positive, the reaction is spontaneous. If the $E°$ value is negative, an outside energy source is necessary for the reaction to occur. In the above example, the $E°$ is a positive 1.10 V, therefore this reaction is spontaneous.

COMPETENCY 0020   UNDERSTAND THE NATURE OF ORGANIC REACTIONS.

Organic reactions can be grouped into basic classes of addition, substitution, elimination, and polymerization along with combustion and oxidation reactions.

An **addition reaction** is a reaction in which two atoms or ions react with a double bond of an alkene, forming a compound with two new functional groups bonded to the carbons of the original double bond. In these reactions, the existing pi bond is broken and in its place, sigma bonds form to two new atoms.

Examples of Simple Addition Reactions

$\diagdown C = C \diagup$ + H—Cl ⟶	(product)	hydrochlorination
$\diagdown C = C \diagup$ + H_2O ⟶	(product)	hydration
$\diagdown C = C \diagup$ + H_2 ⟶	(product)	hydrogenation

The mechanisms for these types of reactions depend upon nucleophilicity/electrophilicity of both the alkene and adding group, and the presence of solvent or catalyst. In many cases, the first step of the reaction is the attack of a positively charged proton by the pi electrons, causing a proton transfer across the double bond. This is followed by addition of the nucleophile to the remaining cation.

Mechanism for Electrophilic Addition of HCl to 2-Butene

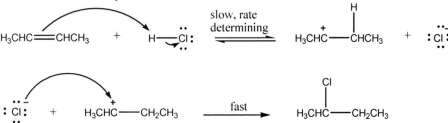

A **substitution reaction** is a reaction in which an atom or group of atoms is replaced by another atom or group of atoms. The most common types of these reactions are S_N1 and S_N2 reactions.

Examples of Substitution Reactions

On of the most important types of substitution reaction is a nucleophilic substitution, an S_N2 reaction in which a halide is added to a molecule. These reactions can lead to a variety of new functional groups. The mechanism for these reactions involves the attack of a nucleophile on a central carbon atom. Simultaneously, β-elimination of a leaving group occurs.

Mechanism for Nucleophilic Substitution of OH⁻ to Bromomethane (S_N2 Reaction)

transistion state with
bond breaking and forming

The mechanism for an S_N1 reaction is a multi-step mechanism, where the leaving group is eliminated in the first step to leave a positively charged carbocation (an electrophile). The cation is then attacked by a nucleophile, followed by the final step of proton transfer to afford a neutral molecule.

An **elimination reaction** is a reaction in which a functional group is split (eliminated) from adjacent carbons. This is a reaction that can often compete with nucleophilic substitution and is highly dependant on the leaving group present in the molecule. Elimination reactions are favored by the presence of strong bases.

Examples of Elimination Reactions

There are two types of elimination reactions, E1 and E2 reactions. The mechanism for E1 is a multistep reaction that involves the formation of a carbocation intermediate. The E2 mechanism is a series of steps, bond breaking and bond formation, that occurs simultaneously. Similar to the S_N2 case outlined above, both the haloalkane and the base are involved in the transition state.

Mechanism for Elimination of 2-Bromo-2-methylpropane (E1 Reaction)

Examples of Polymerization Reactions

Polymers are substances whose molecules have high molar masses and are composed of a large number of repeating units called monomers. For example, polyethylene is many ethylene molecules put together, much like you would link paper clips together to make a long chain.

$$-[CH_2CH_2CH_2CH_2CH_2CH_2CH_2CH_2]_n$$

Polymers are made from the repeating units through two process. One process simply adds each succeeding monomer onto the chain through various methods, called addition polymerization. Condensation polymerization is the second process of adding monomers onto the chain to make large molecules. In condensation polymerization, small molecules like water are removed as the monomer unit is added onto the chain.

It isn't difficult to form addition polymers from monomers containing C=C double bonds; many of these compounds polymerize spontaneously unless polymerization is actively inhibited.

Ethylene has two carbon atoms and four hydrogen atoms, and the polyethylene repeat structure has two carbon atoms and four hydrogen atoms. None gained, none lost.

The simplest way to catalyze the polymerization reaction that leads to an addition polymer is to add a source of a **free radical** to the monomer. The term *free radical* is used to describe a family of very reactive, short-lived components of a reaction that contain one or more unpaired electrons. In the presence of a free radical, addition polymers form by a chain-reaction mechanism that contains chain-initiation, chain-propagation, and chain- termination steps.

Chain Initiation

A source of free radicals is needed to initiate the chain reaction. These free radicals are usually produced by decomposing a peroxide such as di-*tert*-butyl peroxide or benzoyl peroxide, shown below. In the presence of either heat or light, these peroxides decompose to form a pair of free radicals that contain an unpaired electron.

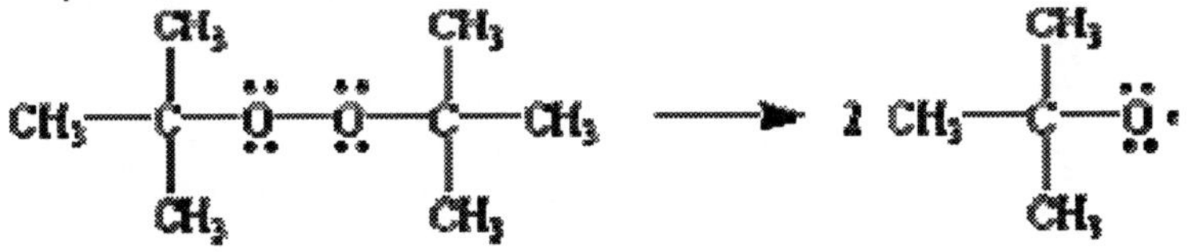

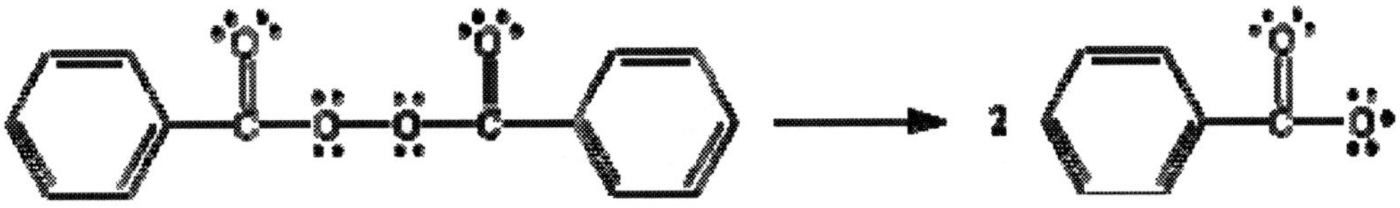

Chain Propagation

The free radical produced in the chain-initiation step adds to an alkene to form a new free radical.

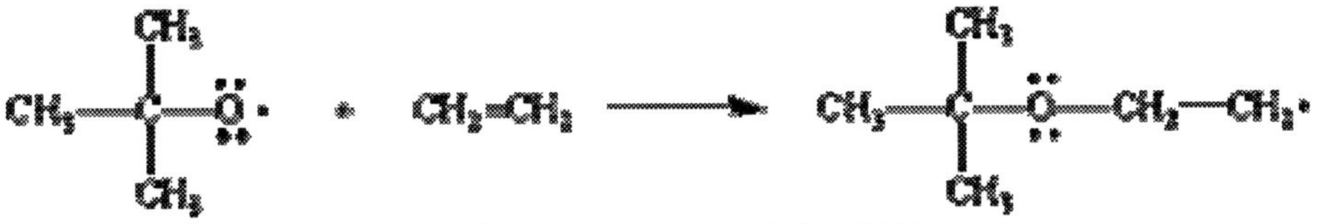

The product of this reaction can then add additional monomers in a chain reaction.

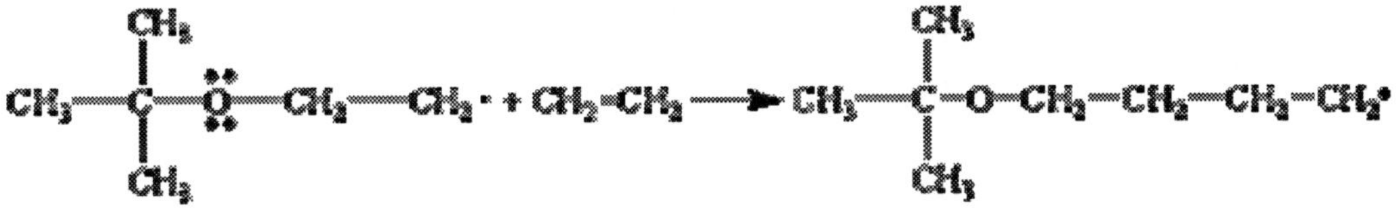

Chain Termination

Whenever pairs of radicals combine to form a covalent bond, the chain reactions carried by these radicals are terminated.

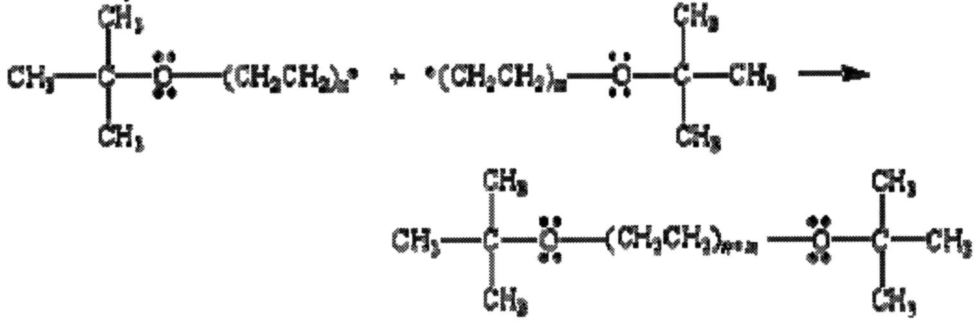

Condensation Polymerization

Monomers involved in condensation polymerization are not like those involved in addition polymerization. These monomers have functional groups like alcohols (-OH), amine (-NH₂) or carboxylic acid (-COOH) instead of double bonds. Each monomer also has at least two active sites at which it can bond.

When an organic acid and an amine react, a water molecule is removed and an amide forms. This is called an amide linkage because of the amide that forms.

When a carboxylic acid and an alcohol react, a water molecule is removed, and an ester molecule is formed. This is referred to as an ester linkage due to the ester that forms.

Mechanism for the formation of polyester

A carboxylic acid monomer and an alcohol monomer can join in an ester linkage.

HO-C(=O)-⟨benzene⟩-C(=O)-OH + H-O-CH$_2$-CH$_2$-OH ⟶ HO-C(=O)-⟨benzene⟩-C(=O)-O-CH$_2$-CH$_2$-OH + H$_2$O

Terephthalic acid   **Ethylene glycol**   **Water**

A water molecule is removed as the ester linkage is formed. Notice the acid and the alcohol groups that are still available for bonding.

HO-C(=O)-⟨benzene⟩-C(=O)-OH

Because the monomers above are all joined by ester linkages, the polymer chain is a polyester. This one is called PET, which stands for poly(ethylene terephthalate). (PET is used to make soft-drink bottles, magnetic tape, and many other plastic products.)

*"Was not all the knowledge
Of the Egyptians writ in
mystic symbols?"*
--Ben Johnson
The Alchemist

COMPETENCY 0021   UNDERSTAND THE MOLE CONCEPT.

A single atom or molecule weighs very little in grams and cannot be measured using a balance in the lab. It's useful to have a system that permits a large number of chemical particles to be described as one unit, analogous to a dozen as 12 of something or one gross as 144. A useful number of atoms, molecules, or formula units is **that number whose mass in grams is numerically equal to the atomic mass, molecular mass, or formula mass** of the substance. This quantity is called the **mole**, abbreviated mol. Because the ^{12}C isotope is assigned an exact value of 12 atomic mass units, there are exactly 12 g of ^{12}C in one mole of ^{12}C. The atomic mass unit is also called a Dalton, and either "u" (for "unified atomic mass unit") or "Da" may be used as an abbreviation. Older texts use "amu." To find the molar mass of a substance use the periodic table at the beginning of this guide to determine the molecular weight of each atom in the substance and multiply by the number of each atom present. For example:

$$Al_2(SO_4)_3 \text{ molecular weight} = 2(26.982 \text{ u for Al}) + 3(32.065 \text{ u for S}) + 12(15.999 \text{ u for O})$$
$$= 342.147 \text{ u.}$$
$$\text{Therefore 1 mol } Al_2(SO_4)_3 = 342.147 \text{ g } Al_2(SO_4)_3.$$

It's been found experimentally that this number of atoms, ions, molecules, or anything else in one mole is 6.022045×10^{23}. For most purposes, three significant digits are sufficient, and **6.02×10^{23}** will be used. This value was named in honor of Amedeo Avogadro after his death and it is referred to as **Avogadro's number**. The following table illustrates why the mole and Avogadro's number are useful. These concepts permit us to think about interactions among individual molecules and atoms while measuring many grams of a substance.

Name	Formula	Formula weight (u)	Mass of 1 mol of formula units (g)	Number and kind of particles in 1 mol
Atomic hydrogen	H	1.0079	1.0079	$6.02 \square 10^{23}$ H atoms
Molecular hydrogen	H_2	2.0158	2.0158	$6.02 \square 10^{23}$ H_2 molecules
				$2(6.02 \square 10^{23})$ H atoms
Silver	Ag	107.87	107.87	$6.02 \square 10^{23}$ Ag atoms
Silver ions	Ag^+	107.87	107.87	$6.02 \square 10^{23}$ Ag^+ ions
Barium chloride	$BaCl_2$	208.24	208.24	$6.02 \square 10^{23}$ $BaCl_2$ units
				$6.02 \square 10^{23}$ Ba^{2+} ions
				$2(6.02 \square 10^{23})$ Cl^- ions

The mole is the chemist's counting unit. Working with the mole should be second nature to students, so let's review grams to mole calculations since they are very important in mass-mass stoichiometry and the balanced equations provides mole ratios not mass ratios.

Mass to moles:

1. First determine the molar mass of the substance by adding the masses for each element in the substance x number of atoms present:

Example: What is the molar mass of $CuSO_4$?
Solution: 1 mole of Cu = 63.5 g + 1 mol of S = 32 g + 4 mol O = 4 x 16 or 48 g = 143.5 g/mol

2. Determine the number of moles present using the molar mass conversion. 1 mol = molar mass of substance for example: 1 mol $CuSO_4$ = 143.5 g

Example: How many moles of $CuSO_4$ are there in 315 g ?

315 g x 1 mol/ 143.5 g =2.20 mol $CuSO_4$

Mole to grams conversions are just the reverse.

Solving these problems is a three-step process:

1) Grams of the given compound are converted to moles.
2) Moles of the given compound are related to moles of a second compound by relating their stoichiometric coefficients from the balanced equation.
3) Moles of the second compound are converted to grams.

These steps are often combined in one series of multiplications, which may be described as **"grams to moles to moles to grams."**

Example: How much mass of oxygen is required to consume 95.0 g of ethane in this reaction: $2C_2H_6 + 7O_2 \rightarrow 4CO_2 + 6H_2O$?

Solution:

$$95.0 \text{ g } C_2H_6 \times \underset{\text{step 1}}{\frac{1 \text{ mol } C_2H_6}{30.1 \text{ g } C_2H_6}} \times \underset{\text{step 2}}{\frac{7 \text{ mol } O_2}{2 \text{ mol } C_2H_6}} \times \underset{\text{step 3}}{\frac{32.0 \text{ g } O_2}{1 \text{ mol } O_2}} = 359 \text{ g } O_2$$

[handwritten annotations: 2×12=24, 6×1=6, 30.1 ✓]

[handwritten: Total grams × $\frac{1 \text{ mole}}{g}$ × $\frac{\text{moles of } O_2}{\text{moles of ethane}}$ × $\frac{\text{grams of } O_2 \text{ per mole}}{1 \text{ mole of } O_2}$ =]

COMPETENCY 0022 UNDERSTAND THE RELATIONSHIP BETWEEN T
CONCEPT OF MOLE AND CHEMICAL FORMULAS.

The expression that shows the number and kind of each atom in a molecule is the chemical formula. Chemical formulas use the chemical symbol of the atom to express the type of atom and a subscript following the symbol to express the number of atoms in the molecule. If there is only one of a type of atom, no subscript is used. In addition to the number and type of atom, chemical formulas can (but don't have to) suggest the arrangement of the atoms.

The chemical formula tells the type and the ratio of atoms in a compound. Because the ratio is the same regardless of whether there is one molecule or millions, we use a "scale-up" factor to work with enough molecules to be easily manipulated. This factor, the SI unit for amount of substance, is a mole. The mole is just a specific number of things, in the case of chemistry either molecules or atoms. Because atoms and molecules are so small, this amount must be very large. The number of particles in a mole is called Avogadro's number and has a value of 6.022×10^{23}. This number is used because the mass of an atom is its atomic weight in atomic mass units, the mass of a mole of atoms is its atomic weight in grams. Since masses are additive, the mass of a molecule is the sum of the mass of each of its atoms in atomic mass units and the mass of a mole of the molecule is the sum of the mass of each of its atoms in grams. The mass of a mole of any substance is called molar mass.

Also, since the ratios are the same in a mole as in a molecule, the molecular formula can be used to obtain mole ratios of elements in a compound.

NaCl represents the compound containing one atom of sodium and one atom of chlorine. The masses on these elements are too small to determine individually or as individual ionic units. However, using the mole we can determine the mass of the compound represented by the formula NaCl. One mole of sodium atoms has a mass of 23.0 grams and one mole of chlorine atoms has a mass of 35.5 grams. Therefore, one mole of sodium chloride molecules has a mass of 58.5 grams.

1. Molar masses of chemical compounds are equal to the sums of the molar masses of all the atoms in one molecule of that compound. If we have a chemical compound like NaCl, the molar mass will be equal to the molar mass of one atom of sodium plus the molar mass of one atom of chlorine. If we write this as a calculation, it looks like this:

(1 atom x 23 grams/mole Na) + (1 atom x 35.5 grams/mole Cl) = 58.5 grams/mole NaCl

subscript in a chemical formula, then you multiply the
s of anything next to that subscript by the number of the
ost compounds, this is easy. For example, in iron (II) chloride,
ve one atom of iron and two atoms of chlorine. The molar mass
1 atom x 56 grams/mole Fe) + (2 atoms x 35.5 grams/mole of
grams/mole of iron (II) chloride.

ounds, this might get a little bit more complicated. For example,
take the example of zinc nitrate, or $Zn(NO_3)_2$. In this compound, we have one
atom of zinc, two atoms of nitrogen (one atom inside the brackets multiplied by
the subscript two) and six atoms of oxygen (three atoms in the brackets
multiplied by the subscript two). The molar mass of zinc nitrate will be equal to:

(1 atom x 65 grams/mole of zinc) + (two atoms x 14 grams/mole of nitrogen) +
(six atoms x 16 grams/mole of oxygen) = 189 grams/mole of zinc nitrate.

Example: Determine the molar mass of potassium hydroxide.

Solution: The chemical formula for potassium hydroxide is KOH.
In potassium hydroxide, there is one atom of potassium, one atom of hydrogen,
and one atom of oxygen. The molar mass will then be (1 mole K x 39 grams/
mol) + (1mol H x 1 gram/ mol) + (1mol O x 16 grams/ mol) = 56 grams/mole of
potassium hydroxide

The **percent composition** of a substance is the **percentage by mass of each
element**. Chemical composition is used to verify the purity of a compound in the
lab. An impurity will make the actual composition vary from the expected one.

To determine percent composition from a formula, do the following:
1) Write down the **number of atoms each element contributes** to the
   formula.
2) Multiply these values by the molecular weight of the corresponding
   element to determine the **grams of each element in one mole** of the
   formula.
3) Add the values from step 2 to obtain the **formula mass**.
4) Divide each value from step 2 by the formula weight from step 3 and
   multiply by 100% to obtain the **percent composition**.

The first three steps are the same as those used to determine formula mass, but
we use the intermediate results to obtain the composition.

Example: What is the chemical composition of ammonium carbonate $(NH_4)_2CO_3$?

Solution:

1) One $(NH_4)_2CO_3$ contains 2 N, 8 H, 1 C, and 3 O.

2) $$\frac{2\ mol\ N}{mol\ (NH_4)CO_3} \times \frac{14.0\ g\ N}{mol\ N} = 28.0\ g\ N/mol\ (NH_4)CO_3$$

$$8(1.0) = 8.0\ g\ H/mol\ (NH_4)CO_3$$
$$1(12.0) = 12.0\ g\ C/mol\ (NH_4)CO_3$$
$$3(16.0) = 48.0\ g\ O/mol\ (NH_4)CO_3$$

Sum is $\overline{96.0\ g\ (NH_4)CO_3/mol\ (NH_4)CO_3}$

3)

4)

$$\%N = \frac{28.0\ g\ N/mol\ (NH_4)_2CO_3}{96.0\ g\ (NH_4)_2CO_3/mol\ (NH_4)_2CO_3} = 0.292\ g\ N/g\ (NH_4)_2CO_3 \times 100\% = 29.2\%$$

$$\%H = \frac{8.0}{96.0} \times 100\% = 8.3\% \quad \%C = \frac{12.0}{96.0} \times 100\% = 12.5\% \quad \%O = \frac{48.0}{96.0} \times 100\% = 50.0\%$$

If we know the chemical composition of a compound, we can calculate the **empirical formula** for it. An empirical formula is the **simplest formula** using the smallest set of integers to express the **ratio of atoms** present in a molecule.

To determine an empirical formula from a percent composition, do the following:
1) Change the "%" sign to grams for a basis of 100 g of the compound.
2) Determine the moles of each element in 100 g of the compound.
3) Divide the values from step 1 by the smallest value to obtain ratios.
4) Multiply by an integer if necessary to get a whole-number ratio.

Example: What is the empirical formula of a compound with a composition of 63.9% Cl, 32.5% C, and 3.6% H?

Solution:

1) We will use a basis of 100 g of the compound containing 63.9 g Cl, 32.5 g C, and 3.6 g H.

2) In 100 g, there are: $63.9\ g\ Cl \times \dfrac{mol\ Cl}{35.45\ g\ Cl} = 1.802\ mol\ Cl$

$$32.5/12.01 = 2.706\ mol\ C$$
$$3.6/1.01 = 3.56\ mol\ H$$

3) Dividing these values by the smallest yields:

$$\frac{2.706 \text{ mol C}}{1.802 \text{ mol Cl}} = 1.502 \text{ mol C/mol Cl}$$

$$\frac{3.56 \text{ mol H}}{1.802 \text{ mol Cl}} = 1.97 \text{ mol H/mol Cl}$$

Therefore, the elements are present in a ratio of C:H:Cl=1.50:2.0:1

4) Multiply the entire ratio by 2 because you cannot have a fraction of an atom. This corresponds to a ratio of 3:4:2 for an empirical formula of $C_3H_4Cl_2$.

The **molecular formula** describing the **actual number of atoms in the molecule** might also be $C_3H_4Cl_2$ or it might be $C_6H_8Cl_4$ or some other multiple that maintains a 3:4:2 ratio.

COMPETENCY 0023   UNDERSTAND QUANTITATIVE RELATIONSHIPS EXPRESSED IN CHEMICAL EQUATIONS.

A properly written chemical equation must contain properly written formulas and must be **balanced**. Chemical equations are written to describe a certain number of moles of reactants becoming a certain number of moles of reaction products. The number of moles of each compound is indicated by its **stoichiometric coefficient**.

Example: In the reaction

$$2H_2(g) + O_2(g) \rightarrow 2H_2O(l),$$

hydrogen has a stoichiometric coefficient of two, oxygen has a coefficient of one, and water has a coefficient of two because 2 moles of hydrogen react with 1 mole of oxygen to form two moles of water.

In a balanced equation, the stoichiometric coefficients are chosen such that the equation contains an **equal number of each type of atom on each side**. In our example, there are four H atoms and two O atoms on both sides. Therefore, the equation is properly written.

Balancing equations is a four-step process.

1) Write an **unbalanced equation**.
2) Determine the **number of each type of atom on each side** of the equation to find if the equation is balanced.
3) Assume that **the molecule with the most atoms** has a stoichiometric coefficient of one, and determine the other stoichiometric coefficients required to create the **same number of atoms on each side** of the equation.
4) Multiply all the stoichiometric coefficients by a whole number if necessary to eliminate fractional coefficients.

Example: Balance the chemical equation describing the combustion of methanol in oxygen to produce only carbon dioxide and water.

Solution:

1) The structural formula of methanol is CH_3OH, so its molecular formula is CH_4O. The formula for carbon dioxide is CO_2. Therefore the unbalanced equation is:

$$CH_4O + O_2 \rightarrow CO_2 + H_2O.$$

2) On the left there are 1C, 4H, and 3O. On the right, there are 1C, 2H, and 3O. It seems close to being balanced, but there's work to do.

3) Assuming that CH_4O has a stoichiometric coefficient of one means that the left side has 1C and 4H that also must be present on the right. Therefore the stoichiometric coefficient of CO_2 will be 1 to balance C and the stoichiometric coefficient of H_2O will be 2 to balance H. Now we have:

$$CH_4O + ?O_2 \rightarrow CO_2 + 2H_2O.$$

and only oxygen remains unbalanced. There are 4O on the right and one of these is accounted for by methanol leaving 3O to be accounted for by O_2.

This gives a stoichiometric coefficient of 3/2 and a balanced equation:

$$CH_4O + \frac{3}{2}O_2 \rightarrow CO_2 + 2H_2O.$$

4) Whole-number coefficients are achieved by multiplying by two:

$$2CH_4O + 3O_2 \rightarrow 2CO_2 + 4H_2O.$$

Reactions among ions in aqueous solution may often be represented in three ways. When solutions of hydrochloric acid and sodium hydroxide are mixed, a reaction occurs and heat is produced. The **molecular equation** for this reaction is:

$$HCl(aq) + NaOH(aq) \rightarrow H_2O(l) + NaCl(aq).$$

It is called a molecular equation because the **complete chemical formulas** of reactants and products are shown. But in reality, both HCl and NaOH are strong electrolytes and exist in solution as ions. This is represented by a **complete ionic equation** that shows all the dissolved ions:

$$H^+(aq) + Cl^-(aq) + Na^+(aq) + OH^-(aq) \rightarrow H_2O(l) + Na^+(aq) + Cl^-(aq).$$

Because $Na^+(aq)$ and $Cl^-(aq)$ appear as both reactants and products, they play no role in the reaction. Ions that appear in identical chemical forms on both sides of an ionic equation are called **spectator ions** because they aren't part of the action. When spectator ions are removed from a complete ionic equation, the result is a **net ionic equation** that shows the actual changes that occur to the chemicals when these two solutions are mixed together:

$$H^+(aq) + OH^-(aq) \rightarrow H_2O(l)$$

Example: Write the balanced molecular and net ionic equation for:

silver nitrate reacts with sodium sulfate producing silver sulfate and sodium nitrate.

Solution: Write the skeleton equation first: $AgNO_3$ (aq) + Na_2SO_4 (aq) → Ag_2S (s) + $NaNO_3$ (aq) Use solubility rules to determine the states of the reactants and products.

Now, balance the molecular equation:
2 $AgNO_3$ (aq) + Na_2SO_4 (aq) → Ag_2SO_4 (s) + 2 $NaNO_3$ (aq)

To write the net ionic equation, break the aqueous species into ions.

2 Ag^+ (aq) + 2 NO_3^- (aq) +2 Na^+ (aq) + SO_4^{2-} (aq) → Ag_2SO_4 (s) +2 Na^+ (aq) +2 NO_3^- (aq) and cancel out species that occur on both the reactant and product side of the equation.

Leaving: 2 Ag^+ (aq) + SO_4^{2-} (aq) → Ag_2SO_4 (s)

An additional requirement for **redox** reactions is that the equation contains an **equal charge on each side**. Redox reactions may be divided into half-reactions which either gain or lose electrons.

There are millions of chemical reactions occurring around us everyday. So many different chemical reactions would be very difficult to understand. However, the millions of chemical reactions that take place each and everyday fall into only a few basic categories. Using these categories can help predict products of reactions that are unfamiliar or new.

nce we have an idea of the **reaction type**, we can make a good prediction
out the products of chemical equations, and also balance the reactions.
eneral reaction types are listed in the following table. Some reaction types
have multiple names.

Reaction type	General equation	Example
Combination	$A + B \rightarrow C$	$2H_2 + O_2 \rightarrow 2H_2O$
Synthesis		
Decomposition	$A \rightarrow B + C$	$2KClO_3 \rightarrow 2KCl + 3O_2$
Single substitution	$A + BC \rightarrow AB + B$	$Mg + 2HCl \rightarrow MgCl_2 + H_2$
Single displacement		
Single replacement		
Double substitution	$AC + BD \rightarrow AD + BC$	$HCl + NaOH \rightarrow NaCl + H_2O$
Double displacement		
Double replacement		
Ion exchange		
Metathesis		
Isomerization	$A \rightarrow A'$	C_3H_6 cyclopropane $\rightarrow$ C_3H_6 propene

Example: Determine the products of a reaction between Cl_2 and a solution of
NaBr.

Solution: The first step is to write "$Cl_2 + NaBr(aq) \rightarrow ?$" Now examine the
possible choices from the table. Decomposition and isomerization
reactions require only one reactant. It also can't be a double
substitution reaction because one of the reactants is an element. A
synthesis reaction to form some NaBrCl compound would require very
unusual valences! The most likely reaction is the remaining one: Cl
replaces Br in aqueous solution with Na:

$Cl_2 + NaBr(aq) \rightarrow NaCl(aq) + Br_2$. After balancing, the equation is:

$Cl_2 + 2NaBr(aq) \rightarrow 2NaCl(aq) + Br_2$.

Many **specific reaction types** also exist. Always determine the complete ionic equation for reactions in solution. This will help you determine the reaction type. The most common specific reaction types are summarized in the following table:

Reaction type	General equation	Example
Precipitation	Molecular: $AC(aq) + BD(aq) \rightarrow AD(s \text{ or } g) + BC(aq)$ Net ionic: $A^+(aq) + D^-(aq) \rightarrow AD(s \text{ or } g)$	Molecular: $NiCl_2(aq) + Na_2S(aq) \rightarrow NiS(s) + 2NaCl(aq)$ Net ionic: $Ni^{2+}(aq) + S^{2-}(aq) \rightarrow NiS(s)$
Acid-base neutralization	Arrhenius: $H^+ + OH^- \rightarrow H_2O$	Arrhenius: $HNO_3 + NaOH \rightarrow NaNO_3 + H_2O$ $H^+ + OH^- \rightarrow H_2O$ (net ionic)
	Brønsted-Lowry: $HA + B \rightarrow HB + A$	Brønsted-Lowry: $HNO_3 + KCN \rightarrow HCN + KNO_3$ $H^+ + CN^- \rightarrow HCN$ (net ionic)
	Lewis: $A + {:}B \rightarrow A{:}B$	Lewis:
Redox	Full reaction: $A + B \rightarrow C + D$	$Ni + CuSO_4 \rightarrow NiSO_4 + Cu$ $Ni + Cu^{2+} \rightarrow Ni^{2+} + Cu$ (net ionic)
	Half reactions: $A \rightarrow C + e^-$ and $e^- + B \rightarrow D$	$Ni \rightarrow Ni^{2+} + 2e^-$ $2e^- + Cu^{2+} \rightarrow Cu$
Combustion	organic molecule $+ O_2 \rightarrow CO_2 + H_2O + $ heat	$2C_2H_6 + 7O_2 \rightarrow 4CO_2 + 6H_2O$

Whether precipitation occurs among a group of ions—and which compound will form the precipitate—may be determined by the solubility rules.

Ion	Solubility	Exception
All compounds containing alkali metals	All are soluble	
All compounds containing ammonium, NH_4^+	All are soluble	
All compounds containing nitrates, NO_3^-	All are soluble	
All compounds containing chlorates, ClO_3^- and perchlorates, ClO_4^-	All are soluble	
All compounds containing acetates, $C_2H_3O_2^-$	All are soluble	
Compounds containing Cl^-, Br^-, and I^-	Are soluble	Except halides of Ag^+, Hg_2^{2+}, and Pb^{2+}
Compounds containing F^-	Are soluble	Except fluorides of Mg^{2+}, Ca^{2+}, Sr^{2+}, Ba^{2+}, and Pb^2
Compounds containing sulfates, SO_4^{2-}	Are soluble	Except sulfates of Mg^{2+}, Ca^{2+}, Sr^{2+}, Ba^{2+}, and Pb^2
All compounds containing carbonates, CO_3^{2-}	Are insoluble	Except those containing alkali metals or NH_4^+
All compounds containing phosphates, PO_4^{3-}	Are insoluble	Except those containing alkali metals or NH_4^+
All compounds containing oxalates, $C_2O_4^{2-}$	Are insoluble	Except those containing alkali metals or NH_4^+
All compounds containing chromates, CrO_4^{2-}	Are insoluble	Except those containing alkali metals or NH_4^+
All compounds containing oxides, O^{2-}	Are insoluble	Except those containing alkali metals or NH_4^+
All compounds containing sulfides, S^{2-}	Are insoluble	Except those containing alkali metals or NH_4^+
All compounds containing sulfites, SO_3^{2-}	Are insoluble	Except those containing alkali metals or NH_4^+
All compounds containing silicates, SiO_4^{2-}	Are insoluble	Except those containing alkali metals or NH_4^+
All compounds containing hydroxides	Are insoluble	Except those containing alkali metals or NH_4^+, or Ba^{2+}

Handwritten notes (left margin): alkali metals, NH_4^+, NO_3^-, ClO_3^-, ClO_4, $C_2H_3O_2$ — yes to soluble

Handwritten notes (left margin): NO — all have exceptions of alkali metals and NH_4^+

If protons are available for combination or substitution then it's likely they are being transferred from an acid to a base. An unshared electron pair on one of the reactants may form a bond in a Lewis acid-base reaction.

The possibility that oxidation numbers may change among the reactants indicates an electron transfer and a redox reaction. Combustion reactants consist of an organic molecule and oxygen.

Example: Determine the products and write a balanced equation for the reaction between sodium iodide and lead(II) nitrate in aqueous solution.

Solution: We know that sodium iodide is NaI and lead(II) nitrate is $Pb(NO_3)_2$.

The reactants of the complete ionic equation are:

$$Na^+(aq) + Pb^{2+}(aq) + I^-(aq) + NO_3^-(aq) \rightarrow ?$$

The solubility rules indicate that lead iodide is insoluble and will form as a precipitate. The unbalanced net ionic equation is then:

$$Pb^{2+}(aq) + I^-(aq) \rightarrow PbI_2(s)$$

and the unbalanced molecular equation is:

$$NaI(aq) + Pb(NO_3)_2(aq) \rightarrow PbI_2(s) + NaNO_3(aq).$$

Balancing yields:

$$2NaI(aq) + Pb(NO_3)_2(aq) \rightarrow PbI_2(s) + 2NaNO_3(aq).$$

Much like a recipe given quantities of ingredients, a balanced chemical equation tells the amounts of reactants needed to produce a given amount of product. These quantities come from the coefficient in the balanced equation.

$NaCl\ (aq) + AgNO_3\ (aq) \rightarrow NaNO_3\ (aq) + AgCl\ (s)$ tells us that 1 mole of sodium chloride solution reacts with 1 mole of silver nitrate solution to produce 1 mole of sodium nitrate solution and 1 mole of solid silver chloride.

The actual masses can be determined by using the molar masses.

Doubling the ingredients in a chocolate chip cookie recipe will yield twice as many cookies. The same is true with chemical equations. Instead of reacting 58.5 grams of NaCl, reacting 117 g of NaCl with 340 g of $AgNO_3$ will produce twice as much sodium nitrate (170 g) and silver chloride (287 g).

Example: Tin (II) fluoride, or stannous fluoride, is used in some dental treatment products. It is made by the reaction of tin with hydrogen fluoride according to the unbalanced equation:

$$Sn(s) + HF (g) \text{ ---> } SnF_2 (s) + H_2 (g)$$

How many grams of SnF_2 are produced from the reaction of 30.00 g of HF with Sn?

Solution: First balance the equation: $Sn(s) + 2 HF (g) \text{ ---> } SnF_2 (s) + H_2 (g)$

Then determine the number of moles given reacting:

$$30.00 \text{ g HF} \times \frac{1mol}{20.0g} = 1.50 \text{ mol HF}$$

Use the balanced equation to convert to mole of asked for(SnF_2):

$$1.50 \text{ mol HF} \times \frac{1molSnF_2}{2molHF} = 0.750 \text{ mol } SnF_2$$

Now, determine the mass of asked for (SnF_2) using the molar mass.

$$0.750 \text{ mol } SnF_2 \times \frac{157.0g}{1mol} = 117.8 \text{ g of } SnF_2 \text{ produced.}$$

Limiting reactants:

In the lab, a reaction is rarely carried out with exactly the required amounts of each reactant. In most cases one or more of the reactants is present in excess; that is, in more than the exact amount required to react with the given amount of the other reactants according to the balanced chemical equation. When all of one reactant is used up, no more product can be formed, even if there is more of the other reactants available. The substance that is completely used up first in a reaction is called the limiting reactant. The limiting reactant controls the amount of product formed in a reaction. The substance that is not used up completely in a reaction is sometimes called the excess reactant.

Example: A chemist only has 6.0 grams of C_2H_2 and an unlimitted supply of oxygen and desires to produce as much CO_2 as possible. If she uses the equation below, how much oxygen should she add to the reaction?

$$2C_2H_2(g) + 5O_2(g) \text{ ---> } 4CO_2(g) + 2 H_2O(l)$$

To solve this problem, it is necessary to determine how much oxygen should be added if all of the reactants were used up (this is the way to produce the maximum amount of CO_2).

First, we calculate the number of moles of C_2H_2 in 6.0 grams of C_2H_2. To be able to calculate the moles we need to look at a periodic table and see that 1 mole of C weighs 12.0 grams and H weighs 1.0 gram. Therefore we know that 1 mole of C_2H_2 weighs 26 grams (2*12 grams + 2*1 gram).

$$6.0\,g\,C_2H_2 \times \frac{1\,mol\,C_2H_2}{(24.0 + 2.0)g\,C_2H_2} = 0.25\,mol\,C_2H_2$$

Then, because there are five (5) molecules of oxygen to every two (2) molecules of C_2H_2, we need to multiply the result by 5/2 to get the total molecules of oxygen. Then we convert to grams to find the amount of oxygen that needs to be added:

$$0.25\,mol\,C_2H_2 \times \frac{5\,mol\,O_2}{2\,mol\,C_2H_2} \times \frac{32.0\,g\,O_2}{1\,mol\,O_2} = 20\,g\,O_2$$

The amounts of products calculated in the stoichiometric problems so far represent theoretical yields. The theoretical yield is the maximum amount of product that can be produced from a given amount of reactant. In most chemical reactions, the amount of product obtained from a reaction is less than the theoretical yield. The measured amount of a product that is obtained from a reaction is called the actual yield of that product.

Chemists strive to vary the conditions of a reaction so that the largest possible yield is obtained. There are many reasons why the actual yield may be less than the theoretical yield. Some of the reactant may take part in side reactions that reduce the amount of desired product, or the product forms in impure form and some product is lost during the purification process.

Chemists are usually interested in the efficiency of a reaction. The efficiency is expressed by comparing the actual yield to the theoretical yield. The percent yield is the ratio of actual yield to theoretical yield multiplied by 100 percent.

% yield = actual yield / theoretical yield x 100

Example: Chlorobenzene is used in the production of many imported chemicals such as aspirin, dyes and disinfectants. One industrial method of producing chlorobenzene is the reaction between benzene (C_6H_6) and chlorine:

$$C_6H_6\,(l) + Cl_2\,(g) \longrightarrow C_6H_5Cl\,(s) + HCl\,(g)$$

When 36.8 grams of benzene react with an excess of Cl_2, the actual yield of C_6H_5Cl was 38.8 g. What was the percent yield?

Solution: % yield= 38.8 or yield = 38%

Theoretical yield is determined as follows:

36.8 g x (1 mol/78.8 g) x (1 mol C_6H_5Cl/ 1 mol C_6H_6) x 112.5 g/mol C_6H_5Cl=52.54 g C_6H_5Cl produced.

% yield= 38.8 g / 52.54 g x 100% =73.85%

ENCY 0024   UNDERSTAND THE PROPERTIES OF SOLUTIONS AND COLLOIDAL SUSPENSIONS, AND ANALYZE FACTORS THAT AFFECT SOLUBILITY.

) or more pure materials mix in a homogeneous way (with their molecules intermixing on a molecular level), the mixture is called a **solution**. Heterogeneous combinations of materials are called **mixtures**. Dispersions of small particles that are larger than molecules are called **colloids**. Liquid solutions are the most common, but any two phases may form a solution. When a pure liquid and a gas or solid form a liquid solution, the pure liquid is called the **solvent** and the non-liquids are called **solutes**. When all components in the solution were originally liquids, then the one present in the greatest amount is called the solvent and the others are called solutes. Solutions with water as the solvent are called **aqueous** solutions. The amount of solute in a solvent is called its **concentration**. A solution with a small concentration of solute is called **dilute**, and a solution with a large concentration of solute is called **concentrated**.

Particles in solution are free to move about and collide with each other, vastly increasing the likelihood that a reaction will occur compared with particles in a solid phase. Aqueous solutions may react to produce an insoluble substance that will fall out of solution as a solid or gas **precipitate** in a **precipitation reaction**. Aqueous solution may also react to form **additional water**, or a different chemical in aqueous solution.

Solubility rules for ionic compounds

Given a cation and anion in aqueous solution, we can determine if a precipitate will form according to some common **solubility rules**.

1) Salts with NH_4^+ or with a cation from group 1 of the periodic table are soluble in water.
2) Nitrates (NO_3^-), acetates ($C_2H_3O_2^-$), chlorates (ClO_3^-), and perchlorates are soluble.
3) Cl^-, Br^-, and I^- salts are soluble except with Ag^+, Hg_2^{2+}, and Pb^{2+}.
4) Sulfates (SO_4^{2-}) are soluble except with Ca^{2+}, Ba^{2+}, Ag^+, Hg_2^{2+}, and Pb^{2+}.
5) Hydroxides (OH^-) are <u>insoluble</u> except with cations from rule 1 and with Ca^{2+}, Sr^{2+}, and Ba^{2+}.
6) Sulfides (S^{2-}), sulfites (SO_3^{2-}), phosphates (PO_4^{3-}), and carbonates (CO_3^{2-}) are insoluble except with cations from rule 1.

The **molarity** (abbreviated M) of a solute in solution is defined as the number of moles of solute in a liter of solution.

$$\text{Molarity} = \frac{\text{moles solute}}{\text{volume of solution in liters}}$$

Molarity is the most frequently used concentration unit in chemical reactions because it reflects the number of solute moles available. By using Avogadro's number, the number of molecules in a flask--a difficult image to conceptualize in the lab--is expressed in terms of the volume of liquid in the flask—a straightforward image to visualize and actually manipulate. Molarity is useful for dilutions because the moles of solute remain unchanged if more solvent is added to the solution:

$$(\text{Initial molarity})(\text{Initial volume}) = (\text{molarity after dilution})(\text{final volume})$$

or

$$M_{initial}V_{initial} = M_{final}V_{final}$$

Example: For example, when 29.25 grams of sodium chloride dissolves in enough water to make one liter of solution.

Solution: the concentration is determined as follows:

$$n = \frac{29.25 \text{ g NaCl}}{58.5 \text{ g/mol}} = 0.500 \text{ moles of NaCl}$$

$$M = \frac{\# \text{ moles of solute}}{\text{Liter solution}} = \frac{0.500 \text{ moles of NaCl}}{1.00 \text{ L solution}} = 0.500 \text{ M NaCl solution}$$

Mass percentage is frequently used to represent every component of a solution (possibly including the solvent) as a portion of the whole in terms of mass.

$$\text{Mass percentage of a component} = \frac{\text{mass of component in solution}}{\text{total mass of solution}} \times 100\%.$$

Example: What is the percent by mass of a solution prepared by dissolving 4.0 g of CH_3COOH is 35.0 g of water?

Solution: % Mass = 4.0 g CH_3COOH / 4.0 g CH_3COOH + 35.0 g H_2O = 4.0 g/40.0 g = 10.%

Parts per million (or ppm) in solution usually refers to a dilute component of a solution as a portion of the whole in terms of mass. A solute present in one part per million would amount to one gram in one million grams of solution. This is also one mg of solute in one kg of solution.

$$\text{Parts per million of a component} = \frac{\text{mass of solute}}{\text{total mass of solution}} \times 10^6$$
$$= \frac{\text{number of mg of solute}}{\text{number of kg of solution}}$$

Strictly speaking, the expression above is a **"ppm by mass."** Parts per million is also sometimes used for ratios of moles or volumes.

A **mole fraction** is used to represent a component in a solution as a portion of the entire number of moles present. If you were able to pick out a molecule at random from a solution, the mole fraction of a component represents the probability that the molecule you picked would be that particular component. Mole fractions for all components must sum to one, and mole fractions are just numbers with no units.

$$\text{Mole fraction of a component} = \frac{\text{moles of component}}{\text{total moles of all components}}$$

The **percent composition** of a substance is the **percentage by mass of each element**. Chemical composition is used to verify the purity of a compound in the lab. An impurity will make the actual composition vary from the expected one.

To determine percent composition from a formula, do the following:
1) Write down the **number of atoms each element contributes** to the formula.
2) Multiply these values by the molecular weight of the corresponding element to determine the **grams of each element in one mole** of the formula.
3) Add the values from step 2 to obtain the **formula mass**.
4) Divide each value from step 2 by the formula weight from step 3 and multiply by 100% to obtain the **percent composition**.

Example: What is the chemical composition of sodium chloride (NaCl)?

Solution:

1) One mol NaCl contains 1 Na and 1 Cl

2) 1 mol Na = 22.99g.
   1 mol Cl = 35.45g.

The sum of na and Cl = 22.99g. + 35.45g. = 58.44g
1 mol of NaCl has 58.44g.

3) The % of Na in NaCl = 22.99/58.44 = 0.39339 times 100 = 39.34
   The % of Cl in NaCl = 35.45/58.44 = 0.6066 times 100 = 60.66

4) % Na in NaCl = 39.34
   % Cl in NaCl = 60.66

As more solid solute particles (circles in the figure to the right) dissolve in a liquid solvent (grey background), the concentration of solute increases, and the chance that dissolved solute (grey circles) will collide with the remaining undissolved solid (white circles) also increases. A collision may result in the solute particle reattaching themselves to the solid. This process is called **crystallization**, and is the

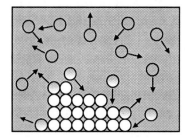

opposite of the solution process. Particles in the act of dissolving or crystallizing are half-shaded in the figure. An animation of the solution process may be found here:

http://www.mhhe.com/physsci/chemistry/essentialchemistry/flash/molvie1.swf.

Equilibrium occurs when no additional solute will dissolve because the rates of crystallization and solution are equal.

$$\text{Solute} + \text{Solvent} \underset{\text{crystallize}}{\overset{\text{dissolve}}{\rightleftharpoons}} \text{Solution}$$

A solution at equilibrium with undissolved solute is a **saturated** solution. The amount of solute required to form a saturated solution in a given amount of solvent is called the **solubility** of that solute. If less solute is present, the solution is called **unsaturated**. It is also possible to have more solute than the equilibrium amount, resulting in a solution that is termed **supersaturated**.

Pairs of liquids that mix in all proportions are called **miscible**. Liquids that don't mix are called **immiscible**.

Intermolecular forces in the solution process

Solutions tend to form when the intermolecular attractive forces between solute and solvent molecules are about as strong as those that exist in the solute alone or in solvent alone. NaCl dissolves in water because:

1) The water molecules interact with the Na^+ and Cl^- ions with sufficient strength to overcome the attraction between them in the crystal.

2) Na^+ and Cl^- ions interact with the water molecules with sufficient strength to overcome the attraction water molecules have for each other in the liquid.

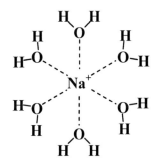

The intermolecular attraction between solute and solvent molecules is known as **solvation**. When the solvent is water, it is known as **hydration**. The figure to the left shows a hydrated Na^+ ion.

Polar and nonpolar solutes and solvents

A nonpolar liquid like heptane (C_7H_{16}) has intermolecular bonds with relatively weak London dispersion forces. Heptane is immiscible in water because the attraction that water molecules have for each other via hydrogen bonding is too strong. Unlike Na^+ and Cl^- ions, heptane molecules cannot break these bonds. Because bonds of similar strength must be broken and formed for solvation to occur, nonpolar substances tend to be soluble in nonpolar solvents, and ionic and polar substances are soluble in polar solvents like water. Polar molecules are often called **hydrophilic** and non-polar molecules are called **hydrophobic**. This observation is often stated as "**like dissolves like.**" Network solids (e.g., diamond) are soluble in neither polar nor nonpolar solvents because the covalent bonds within the solid are too strong for these solvents to break.

Electrolytes

NaCl is present in solution as ions. Compounds that are completely ionized in water are called **strong electrolytes** because these solutions easily conduct electricity. Most salts are strong electrolytes. Other compounds (including many acids and bases) may dissolve in water without completely ionizing. These are referred to as **weak electrolytes** and their state of ionization is at equilibrium with the larger molecule. Those compounds that dissolve with no ionization (e.g., glucose, $C_6H_{12}O_6$) are called **nonelectrolytes**.

Gas solubility

Pressure does not dramatically alter the solubility of solids or liquids, but kinetic molecular theory predicts that **increasing the partial pressure of a gas will increase the solubility of the gas** in a liquid. If a substance is distributed between gas and solution phases and pressure is exerted, more gas molecules will impact the gas/liquid interface per second, so more will dissolve until a new equilibrium is reached at a higher solubility. **Henry's law** describes this relationship as a direct proportionality:

$$\text{Solubility of gas in liquid (in } \frac{\text{mol solute}}{\text{L solution}}) \propto P_{gas}$$

Carbonated drinks are bottled under high CO_2 pressure, permitting the gas to dissolve into aqueous solution. When the bottle is opened, the partial pressure of CO_2 in the gas phase rapidly decreases to the value in the atmosphere, and the gas bubbles out of solution. When the bottle is closed again, CO_2 gas pressure builds until a saturated solution at equilibrium is again obtained. The solubility of gases in a liquid also increases in the bloodstream of deep-sea divers when they experience high pressures. If they return to atmospheric pressure too rapidly, large bubbles of nitrogen gas will form in their blood and cause a potentially lethal condition known as **the bends** or **decompression sickness**. The diver must enter a hyperbaric (high pressure) chambers to redissolve the nitrogen back into the blood.

Increasing temperature will decrease the solubility of a gas in a liquid because kinetic energy opposes intermolecular attractions and permits more molecules to escape from the liquid phase. The vapor pressure of a pure liquid increases with temperature for the same reason. Greater kinetic energy will favor material in the gas phase.

Liquid and solid solubility
For solid and liquid solutes, the impact of temperature depends on whether the solution process requires or releases heat. The following brief analysis is applicable for the effect of temperature on solutions.

Three processes occur when a solution is formed:
1) Solute particles are separated from each other, and heat is required to break these bonds.
2) Solvent particles are separated from each other to create space for solute particles, and heat is required to break these bonds also.
3) Solute and solvent particles interact with each other forming new bonds, and releasing heat.

If the heat required for the first two processes is greater than the heat released by the third, then the entire reaction may be written as an endothermic process:

$$Solute + Solvent + Heat \rightleftharpoons Solution$$

and according to Le Chatelier's principle, **solubility will increase with increasing temperature for an endothermic solution process**. This occurs for most salts in water, including NaCl. There is a large increase for potassium nitrate—KNO_3.

However, heat is released when many solutes enter solution, and the entire reaction is exothermic:

$$Solute + Solvent \rightleftharpoons Solution + Heat.$$

Solubility will decrease with increasing temperature for an exothermic solution process. This is the case for cerium(III) sulfate—$Ce_2(SO_4)_3$—in water.

Summary
The following table summarizes the impact of temperature and pressure on solubility:

Effect on solution of an **increase** in one variable with the other constant	− = decrease, **0** = no/small change, **+** = increase, **++** = strong increase				
	Gas solute in liquid solvent			Solid and liquid solutes	
	Average kinetic energy of molecules	Collisions of gas with liquid interface	Solubility	Solubility for an endothermic heat of solution	Solubility for an exothermic heat of solution
Pressure	0	++	+	0	0
Temperature	+	+	−	+	−

Impact of surface area

It is apparent that dissolution will only take place at the surface of a solute where the solvent molecules contact the solute. When this surface area is increased, the solute dissolves more rapidly. **The rate of solution is increased by surface area, but the solubility remains unchanged.** Solubility is the amount of solute that will saturate a solution, and this depends only on pressure, temperature, and the identity of the solute and solvent.

Granulated sugar, for example, will dissolve more rapidly in water than a sugar cube. This is because it takes very little time for water to dissolve through relatively thin layers of sugar in a small grain before reaching the center of each grain. This is in contrast to the sugar in the center of a sugar cube. The sugar in the center of a cube cannot dissolve for until the thicker layers of sugar around it has dissolved away into solution. However, the solubility of sugar in water does not depend on whether sugar cubes or granulated sugar is used.

Colligative Properties

Do you add a handful of salt to the pot of water when you cook spaghetti? Do you add rock salt to the ice when you make home-made ice or sprinkle calcium chloride on your sidewalks in the winter to keep them from freezing? Do you have antifreeze in your car's radiator? You probably do all of the above (unless you live in a warm climate!). Do you have any idea why these things are done?

By adding a solute to a pure substance (the solvent), some physical properties of the solvent are altered. These properties are called colligative properties. A colligative property depends on how many solute particles are dissolved not the nature of these solute particles.

Two common colligative properties are freezing point depression and boiling point elevation.

Recall that for a liquid to freeze the particles must line up to form a regular repeating pattern or lattice structure with strong attractive forces holding the particles in place. The temperature at which this alignment is achieved is the freezing point temperature.

However, when a solute is present the solvated molecules are not as regularly shaped and are larger so to line up in the same regular repeating pattern of the solid, the molecules must move even slower. Therefore, the temperature at which the alignment is achieved is less than if there was no solute present. The solute, then, depresses the freezing point temperature. Mathematically, this temperature change can be represented:

$$\Delta T = K_f \, m \, i \quad \text{where} \quad K_f \text{ is the freezing point depression constant.}$$

This is a constant value and is unique for each solvent. The K_f for water is -1.86 °C/m. The number of ions present when the solute dissociates is represented by the symbol i and is called the van Hoft factor. Sucrose, being molecular, provides only one particle in solution while sodium chloride dissociates into two ions, Na^+ and Cl^-. Aluminum chloride produces three ions in solution, Al^{3+} and 3 Cl^- ions.

For example, the custard used to make ice cream does not "freeze" until the temperature reaches -15 to -18 °C. With ice, it will only go down to 0 °C. In order to reach the lower temperature needed to freeze the custard, sodium chloride, NaCl, is added to the ice. Each sodium chloride particle dissociates into a Na^+ ion and a Cl^- ion, making two solvated ions present for every one NaCl particle present. So, NaCl has an i value of 2. The sodium and chlorine ions lower the temperature at which the solution will freeze. How much lower, depends on how many ions are present.

The temperature change needed is -15°C and the K_f value for water is -1.86 °C/m.

Using the freezing point depression expression we can determine the *molality* of the salt-ice solution that will reach -15°C.

$\Delta T = K_f \, m \, i$ and rearranging it to solve for $m = \dfrac{\Delta T}{} = \dfrac{-15°C}{} = 4.0 \, m$

Remember that,

$$m = \frac{\text{\# moles solute,}}{\text{kg solvent}}$$

so the moles of NaCl can be determined if, we know the kilograms of ice used. In this case, 1 bag of ice is about 2.3 kg. So the number of moles of NaCl needed would be:

\# moles solute = m (kg solute) = 4.0 *mol/kg* (2.3 kg) = 9.3 mol of NaCl.

Now, the mass of NaCl needed can be found by using the relationship between moles and molar mass: \# mol = mass/ Molar Mass

In this case, mass = \# mol (Molar Mass) = 9.3 mol (58.5 g/mol) or about 540 g of NaCl is needed for every bag of ice used.

A solution will also boil at a higher temperature than the pure solvent. The pure solvent boils when the atmospheric pressure equals the vapor pressure.

When a pure liquid solvent is changed to a vapor, molecules at the surface overcome the attractive forces and escape to become vapor molecules with greater kinetic energies than the liquid molecules. These vapor molecules collide with the surface of the liquid creating a pressure on the liquid surface. The substance boils when the number of collisions between vapor molecules and the liquid's surface equals the number of collisions between air molecules and the container.

The addition of a solute raises the temperature where the number of collisions are equal. The dissolving of a solute created particles that are larger and irregular in size meaning that fewer solvated particles will fit on the surface of the liquid; fewer particles to escape and create pressure. That means the temperature must increase to change enough particles into vapor to create a pressure the same as the atmospheric pressure.

The needed elevation in temperature can be determined by

$$\Delta T_b = K_b \, m \, i \,,$$

where K_b is the boiling point constant and is unique for each solvent. For water K_b equal 0.52 °C/m. The molal concentration of the solute is represented by m and the number of ions present when the solute dissolves in the solvent is shown by i.

For example: Adding salt to a pot of water will increase the temperature at which the water boils.By how much? A 6 quart dutch oven holds about 6 liters of water. A box of salt contains 728 g of sodium chloride.

How much salt is needed to increase the boiling point temperature of the water in the dutch oven by 1.5°C?

Using $\Delta T_b = K_b\, m\, i$ we know: ΔT_b = 1.5°C

i= 2 *for NaCl dissociates into Na^+ and Cl^- ions*
K_b = 0.52 °C/m

To solve for *m* we rearrange : $m = \dfrac{\Delta T_b}{K_b\, i}$ or $m = \dfrac{1.5\ °C}{0.52\ °C/m\ (2)}$ = 1.44 *m*

$m = \dfrac{\#\ moles\ solute}{Kg\ solvent}$ = 1.44 = $\dfrac{\#\ moles\ solute}{6\ kg\ water}$ = 8.65 moles of salt must be added

SUBAREA VI.   INTERACTIONS OF CHEMISTRY, SOCIETY, AND THE ENVIRONMENT

COMPETENCY 0025   UNDERSTAND THE HISTORICAL AND CONTEMPORARY CONTEXTS OF THE STUDY OF CHEMISTRY.

Development of Modern Chemistry

Chemistry emerged from two ancient roots: craft traditions and philosophy. The oldest ceramic crafts (i.e., pottery) known are from roughly 10000 BC in Japan. Metallurgical crafts in Eurasia and Africa began to develop by trial and error around 4000-2500 BC resulting in the production of copper, bronze, iron, and steel tools. Other craft traditions in brewing, tanning, and dyeing led to many useful empirical ways to manipulate matter.

Ancient philosophers in Greece, India, China, and Japan speculated that all matter was composed of four or five elements. The Greeks thought that these were: fire, air, earth, and water. Indian philosophers and Aristotle from Greece also thought a fifth element—"aether" or "quintessence"—filled all of empty space. The Greek philosopher Democritus thought that matter was composed of indivisible and indestructible atoms. These concepts are now known as classical elements and classical atomic theory.

Before the emergence of the scientific method, attempts to understand matter relied on alchemy: a mixture of mysticism, best guesses, and supernatural explanations. Goals of alchemy were the transmutation of other metals into gold and the synthesis of an elixir to cure all diseases. Ancient Egyptian alchemists developed cement and glass. Chinese alchemists developed gunpowder in the 800s AD.

During the height of European alchemy in the 1300s, the philosopher, William of Occam proposed the idea that when trying to explain a process or develop a theory, the simplest explanation with the fewest variables is best. This is known as Occam's Razor. European alchemy slowly developed into modern chemistry during the 1600s and 1700s. This began to occur after Francis Bacon and René Descartes described the scientific method in the early 1600s.

Robert Boyle was educated in alchemy in the mid-1600s, but he published a book called *The Skeptical Chemist* that attacked alchemy and advocated using the scientific method. He is sometimes called the founder of modern chemistry because of his emphasis on proving a theory before accepting it, but the birth of modern chemistry is usually attributed to Lavoisier. Boyle rejected the 4 classical elements and proposed the modern definition of an element. Boyle's law states that gas volume is proportional to the reciprocal of pressure.

Blaise Pascal in the mid-1600s determined the relationship between pressure and the height of a liquid in a barometer. He also helped to establish the scientific method. The SI unit of pressure is named after him.

Isaac Newton studied the nature of light, the laws of gravity, and the laws of motion around 1700. The SI unit of force is named after him.

Daniel Bernoulli proposed the kinetic molecular theory for gases in the early 1700s to explain the nature of heat and Boyle's Law. At that time, heat was thought to be related to the release of a substance called *phlogiston* from combustible material.

James Watt created an efficient steam engine in the 1760s-1780s. Later chemists and physicists would develop the theory behind this empirical engineering accomplishment. The SI unit of power is named after him.

Joseph Priestley studied various gases in the 1770s. He was the first to produce and drink carbonated water, and he was the first to isolate oxygen from air. Priestley thought oxygen was air with its normal phlogiston removed so it could burn more fuel and accept more phlogiston than natural air.

Antoine Lavoisier is called the father of modern chemistry because he performed quantitative, controlled experiments. He carefully weighed material before and after combustion to determine that burning objects gain weight. Lavoisier formulated the rule that chemical reactions do not alter total mass after finding that reactions in a closed container do not change weight. This disproved the plogiston theory, and he named Priestley's substance oxygen. He demonstrated that air and water were not elements. He defined an element as a substance that could not be broken down further. He published the first modern chemistry textbook, *Elementary Treatise of Chemistry*. Lavoisier was executed in the Reign of Terror at the height of the French Revolution.

Additional Gas Laws in the 1700s and 1800s
These contributions built on the foundation developed by Boyle in the 1600s.

Jacque Charles developed Charles's law in the late 1700s. This states that gas volume is proportional to absolute temperature.

William Henry developed the law stating that solubility of gas in a liquid is proportional to the pressure of gas over the liquid. This is known as Henry's Law

Joseph Louis Gay-Lussac developed the gas law stating that gas pressure is directly proportional to absolute temperature. He also determined that 2 volumes of hydrogen react with one of oxygen to produce water and that other reactions occurred with similar simple ratios. These observations led him to develop the Law of Combining Volumes.

Amedeo Avogadro developed the hypothesis that equal volumes of different gases contain an equal number of molecules, if the gases are at the same temperature and pressure. The proportionality between volume and number of moles is called Avagadro's Law, and the number of molecules in a mole is called Avagadro's Number. Both were posthumously named in his honor.

Thomas Graham developed Graham's Law of effusion and diffusion in the 1830s. He is called the father of colloid chemistry.

Electricity and Magnetism in the 1700s and 1800s
Benjamin Franklin studied electricity in the mid-1700s. He developed the concept of positive and negative electrical charges. His most famous experiment showed that lightning is an electrical process.

Luigi Galvani discovered bioelectricity. In the late 1700s, he noticed that the legs of dead frogs twitched when they came into contact with an electrical source.

In the late 1700s, Charles Augustin Coulomb derived mathematical equations for attracti**on and repulsion** between electrically charged objects.

Alessandro **Volta** built the first **battery** in 1800 permitting future research and applications to have a source of continuous electrical current available. The SI unit of electric potential difference is named after him.

André-Marie **Ampère** created a mathematical theory in the 1820s for magnetic fields and electric currents. The SI unit of electrical current is named after him.

Michael **Faraday** is best known for work in the 1820s and 1830s establishing that a moving magnetic field induces an electric potential. He built the first **dynamo** for electricity generation. He also discovered benzene, invented oxidation numbers, and popularized the terms *electrode*, *anode*, and *cathode*. The SI unit of electrical capacitance is named in his honor.

James Clerk **Maxwell** derived the **Maxwell Equations** in 1864. These expressions completely describe **electric and magnetic fields** and their interaction with matter. Also see Ludwig Boltzmann below for Maxwell's contribution to thermodynamics.

Nineteenth Century Chemistry: Caloric Theory and Thermodynamics
Lavoisier proposed in the late 18th century that the heat generated by combustion was due to a weightless material substance called **caloric** that flowed from one place to another and was never destroyed.

In 1798, **Benjamin Thomson**, also known as **Count Rumford** measured the heat produced when cannon were bored underwater and concluded that caloric was not a conserved substance because heat could continue to be generated indefinitely by this process.

Sadi **Carnot** in the 1820s used caloric theory in developing theories for the **heat engine** to explain the engine already developed by Watt. Heat engines perform mechanical work by expanding and contracting a piston at two different temperatures.

In the 1820s, Robert **Brown** observed dust particles and particles in pollen grains moving in a random motion. This was later called **Brownian motion**.

Germain Henri **Hess** developed **Hess's Law** in 1840 after studying the heat required or emitted from reactions composed of several steps.

James Prescott **Joule** determined the equivalence of heat energy to mechanical work in the 1840s by carefully measuring the heat produced by friction. Joule attacked the caloric theory and played a major role in the acceptance of **kinetic molecular theory**. The SI unit of energy is named after him.

William Thomson, 1st Baron of Kelvin also called **Lord Kelvin** recognized the existence of **absolute temperature** in the 1840s and proposed the temperature scale named after him. He failed in an attempt to reconcile caloric theory with Joule's discovery and caloric theory began to fall out of favor.

Hermann von **Helmholtz** in the 1840s proposed that **energy is conserved** during physical and chemical processes, not heat as proposed in caloric theory

Rudolf **Clausius** in the 1860s introduced the concept of **entropy**.

In the 1870s, Ludwig **Boltzmann** generalized earlier work by Maxwell solving the **velocity or energy distribution among gas molecules**. The final diagram in shows the Maxwell-Boltzmann distribution for kinetic energy at two temperatures. Maxwell's contribution to electromagnetism is described above.

Johannes **van der Waals** in the 1870s was the first to consider **intermolecular attractive forces** in modeling the behavior of liquids and non-ideal gases.

Francois Marie **Raoult** studied colligative properties in the 1870s. He developed **Raoult's Law** relating solute and solvent mole fraction to vapor pressure lowering.

Jacobus **van't Hoff** was the first to fully describe **stereoisomerism** in the 1870s. He later studied **colligative properties** and the impact of temperature on equilibria.

Josiah Willard **Gibbs** studied thermodynamics and statistical mechanics in the 1870s. He formulated the concept now called **Gibbs free energy** that will determine whether or not a chemical process at constant pressure will spontaneously occur.

Henri Louis **Le Chatelier** described chemical **equilibrium** in the 1880s using **Le Chatelier's Principle**.

In the 1880s, Svante **Arrhenius** developed the idea of **activation energy.** He also described the dissociation of salts—including **acids and bases**—into ions. Before then, salts in solution were thought to exist as intact molecules and ions were mostly thought to exist as electrolysis products. Arrhenius also predicted that CO_2 emissions would lead to global warming .

In 1905, **Albert Einstein** created a **mathematical model of Brownian motion** based on the impact of water molecules on suspended particles. Kinetic molecular theory could now be observed under the microscope. Einstein's more famous later work in physics on **relativity** may be applied to chemistry by correlating the energy change of a chemical reaction with extremely small changes in the total mass of reactants and products.

Nineteenth and Twentieth Century: Atomic Theory

See section 0006 for the contributions to atomic theory of John **Dalton, J. J. Thomson**, Max **Planck,** Ernest **Rutherford,** Niels **Bohr,** Louis **de Broglie, **Werner **Heisenberg,** and Erwin **Schrödinger**.

Wolfgang **Pauli** helped to develop quantum mechanics in the 1920s by forming the concept of spin and the **exclusion principle.**

Friedrich **Hund** determined a set of **rules to determine the ground state** of a multi-electron atom in the 1920s. One particular rule is called **Hund's Rule** in introductory chemistry courses.

Discovery and Synthesis: Nineteenth Century:

Humphry **Davy** used Volta's battery in the early 1800s for **electrolysis of salt solutions**. He synthesized several pure elements using electrolysis to generate non-spontaneous reactions.

Jöns Jakob **Berzelius** isolated several elements, but he is best known for inventing modern **chemical notation** by using one or two letters to represent elements in the early 1800s.

Friedrich **Wöhler** isolated several elements, but he is best known for the chemical **synthesis of an organic compound** in 1828 using the carbon in silver cyanide. Before Wöhler, many had believed that a transcendent "life-force" was needed to make the molecules of life.

Justus **von Liebig** studied the chemicals involved in agriculture in the 1840s. He has been called the **father of agricultural chemistry**.

Louis **Pasteur** studied **chirality** in the 1840s by separating a mixture of two chiral molecules. His greater contribution was in biology for discovering the germ theory of disease.

Henry **Bessemer** in the 1850s developed the **Bessemer Process** for mass producing steel by blowing air through molten iron to oxidize impurities.

Friedrich August **Kekulé** von Stradonitz studied the chemistry of carbon in the 1850s and 1860s. He proposed the **ring structure of benzene** and that carbon was tetravalent.

Anders Jonas **Ångström** was one of the founders of the science of spectroscopy. In the 1860s, he found hydrogen and other **elements in the spectrum of the sun**. A non-SI unit of length equal to 0.1 nm is named for him.

Alfred **Nobel** invented the explosive **dynamite** in the 1860s and continued to develop other explosives. In his will he used his fortune to establish the **Nobel Prizes**.

Dmitri **Mendeleev** developed the first modern **periodic table** in 1869.

Discovery and Synthesis: Turn of the 20th Century

William **Ramsay** and Lord **Rayleigh** (John William Strutt) isolated the **noble gases**.

Wilhelm Konrad **Röntgen** discovered **X-rays**.

Antoine Henri **Becquerel discovered radioactivity** using uranium salts.

Marie **Curie** named the property radioactivity and determined that it was **a property of atoms** that did not depend on which molecule contained the element.

Pierre and Marie **Curie** utilized the properties of radioactivity to **isolate radium** and other radioactive elements. Marie Curie was the first woman to receive a Nobel Prize and the first person to receive two. Her story continues to inspire. See http://nobelprize.org/physics/articles/curie/index.html for a biography.

Frederick **Soddy** and William **Ramsay** discovered that **radioactive decay can produce helium** (alpha particles).

Fritz **Haber** developed the **Haber Process** for synthesizing ammonia from hydrogen and nitrogen using an iron **catalyst**. Ammonia is still produced by this method to make fertilizers, textiles, and other products.

Robert Andrew **Millikan** determined the **charge of an electron** using an oil-drop experiment.

Discovery and Synthesis: 20th Century

Gilbert Newton **Lewis** described **covalent bonds** as sharing electrons in the 1910s and the **electron pair donor/acceptor theory of acids and bases** in the 1920s. Lewis dot structures and Lewis acids are named after him.

Johannes Nicolaus **Brønsted** and Thomas Martin **Lowry** simultaneously developed the **proton donor/acceptor theory of acids and bases** in the 1920s.

Irving **Langmuir** in the 1920s developed the science of **surface chemistry** to describe interactions at the interface of two phases. This field is important to heterogeneous catalysis .

Fritz **London** studied the electrical nature of chemical bonding in the 1920s. The weak intermolecular **London dispersion forces** are named after him.

Hans Wilhelm **Geiger** developed the **Geiger counter** for measuring ionizing radiation in the 1930s.

Wallace **Carothers** and his team first synthesized **organic polymers** (including neoprene, polyester and nylon) in the 1930s.

In the 1930s, Linus **Pauling** published his results on **the nature of the covalent bond.** Pauling electronegativity is named after him. In the 1950s, Pauling determined the α-helical structure of proteins.

Lise **Meitner** and Otto **Hahn** discovered **nuclear fission** in the 1930s.
Glenn Theodore **Seaborg** created and isolated several **elements larger than uranium** in the 1940s. Seaborg reorganized the periodic table to its current form.

James **Watson** and Francis **Crick** determined the double helix structure of DNA in the 1950s.

Neil **Bartlett** produced **compounds containing noble gases** in the 1960s, proving that they are not completely chemically inert.

Harold Kroto, Richard Smalley, and Robert Curl discovered the **buckyball C_{60}** in the 1980s.

COMPETENCY 0026 UNDERSTAND THE CHEMISTRY OF PRACTICAL PROCESSES AND APPLICATIONS OF CHEMICAL THEORY TO OTHER SCIENTIFIC DISCIPLINES.

Chemistry is referred to as the "central science" and it is with a good reason. Many experiences in our everyday life involve chemistry. Other scientific disciplines rely on chemical concepts to help explain phenomena and solve problems in their respective disciplines. Many industrial processes that produce the goods we often take for granted make use of chemical concepts in everyday manufacturing.

For example:

- Bread is leavened by the use of baking powder; water and a mixture of an acid-producing ingredient, which causes a chemical change to occur as CO_2 gas is produced.
- Casein glue, a wood glue, forms as a protein is denatured by a basic solution.
- Fabric dyes, often extracted from plant pigments, are acid-base indicators.
- Popping corn uses gas laws to explain the expanding popcorn hull.
- Homemade ice cream utilizes colligative properties to lower the freezing point of the ice so that the custard mixture may freeze.
- Making Kool-aid involves solution chemistry with varying concentrations of Kool-aid or speeding up the dissolving process.
- DNA analysis involves a technique called electrophoresis. Electrophoresis separates components of mixtures based on particle size and charge.
- Weathering of rocks is cause by chemical reactions.
- Cellular functions are the result of chemical reactions: photosynthesis, respiration, use of enzymes in digestion to break down complex molecules
- Fractional distillation of petroleum separates petroleum into usable components like gasoline, kerosene, heating oil, etc.
- Polymerization processes to make plastics found in virtually every product.
- Composite materials used in graphite tennis rackets or titanium golf clubs for strength and durability.

COMPETENCY 0027   UNDERSTAND THE APPLICATION OF NUCLEAR REACTIONS.

When two nuclei collide, they sometimes attach to each other and synthesize a new nucleus. This **nuclear fusion** was first demonstrated by the synthesis of oxygen from nitrogen and alpha particles:

$$^{14}_{7}N + ^{4}_{2}He \rightarrow ^{17}_{8}O + ^{1}_{-1}H.$$

Fusion is also used to create new heavy elements, causing periodic tables to grow out of date every few years. In 2004, IUPAC approved the name roentgenium (in honor of Wilhelm Roentgen, the discoverer of X-rays) for the element first synthesized in 1994 by the following reaction:

$$^{209}_{83}Bi + ^{64}_{28}Ni \rightarrow ^{272}_{111}Rg + ^{1}_{0}n.$$

A heavy nucleus may also split into smaller nuclei by **nuclear fission. Nuclear power** currently provides 17% of the world's electricity. Heat is generated by **nuclear fission of uranium-235 or plutonium-239**. This heat is then converted to electricity by boiling water and forcing the steam through a turbine. Fission of ^{235}U and ^{239}Pu occurs when **a neutron strikes the nucleus and breaks it apart into smaller nuclei and additional neutrons**. One possible fission reaction is:

$$^{1}_{0}n + ^{235}_{92}U \rightarrow ^{141}_{56}Ba + ^{92}_{36}Kr + 3\,^{1}_{0}n$$

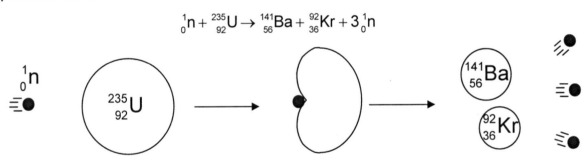

Gamma radiation, kinetic energy from the neutrons themselves, and the decay of the fission products (^{141}Ba and ^{92}Kr in the example above) all produce heat. The neutrons produced by the reaction strike other uranium atoms and produce more neutrons and more energy in a **chain reaction**. If enough neutrons are lost, the chain reaction stops and the process is called **subcritical**. If the mass of uranium is large enough so that one neutron on average from each fission event triggers another fission, the reaction is said to be **critical**. If the mass is larger than this so that few neutrons escape, the reaction is called **supercritical**. The chain reaction then multiplies the number of fissions that occur and the violent explosion of an atomic bomb will take place if the process is not stopped. The concentration of **fissile material** in nuclear power plants is sufficient for a critical reaction to occur but too low for a supercritical reaction to take place.

The alpha decay of **Plutonium-238 is used as a heat source for localized power generation** in space probes and in heart pacemakers from the 1970s.

The most promising nuclear reaction for producing power by nuclear fusion is:

$$_1^2H + _1^3H \rightarrow _2^4He + _0^1n$$

Hydrogen-2 is called **deuterium** and is often represented by the symbol D. Hydrogen-3 is known as **tritium** and is often represented by the symbol T. Nuclear reactions between very light atoms similar to the reaction above are the energy source behind the sun and the hydrogen bomb.

Interconversion of mass and energy

With nuclear reactions, the energies involved are so great that the changes in mass become easily measurable. One no longer can assume that mass and energy are conserved separately, but must take into account their interconversion via Einstein's relationship, **E = mc²**. If mass is in grams and the velocity of light is expressed as **c** = 3×10^{10} cm sec^{-1}, then the energy is in units of g cm² sec^{-2}, or ergs. A useful conversion is from mass in amu to energy in million electron volts (MeV):

1 amu = 931.4 MeV

Which holds a nucleus together? If we attempt to bring two protons and two neutrons together to form a helium nucleus, we might reasonably expect the positively charged protons to repel one another violently. Then what keeps them together in the $_2^4$He nucleus?

The answer is that a helium atom is lighter than the sum of two protons, two neutrons, and two electrons. Some of the mass of the separated particles is converted into energy and dissipates when the nucleus is formed. Before the helium nucleus can be torn apart into its component particles, this dissipated energy must be restored and turned back into mass. Unless this energy is provided, the nucleus cannot be taken apart. This energy is termed the *binding energy* of the helium nucleus.

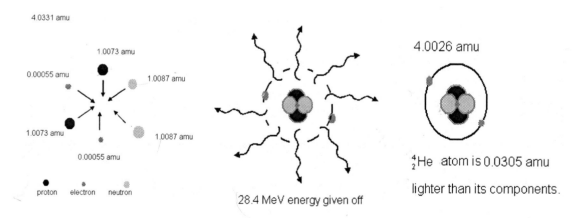

4.0331 amu

1.0073 amu

0.00055 amu
1.0087 amu

1.0073 amu
1.0087 amu

0.00055 amu

proton   electron   neutron

28.4 MeV energy given off

4.0026 amu

$_2^4$He atom is 0.0305 amu lighter than its components.

(We must include electrons in this calculation because 4.0026 amu is the mass of the helium-4 atom, not the nucleus.) This missing mass corresponds to 0.0305 x 931.4 MeV = 28.4 MeV of energy. If we could put together a helium atom directly from two neutrons, two protons, and two electrons, then 28.4 MeV of energy would be given off for every atom formed:

$$2p + 2n + 2e^- \rightarrow {}^4_2He + 28.4 \text{ MeV of energy}$$

Compared to common chemical reactions, this is an enormous quantity of energy. Since 1 electron volt per atom is equivalent to 23.06 kcal per mole,

[handwritten: 1 e⁻ volt = 23.06 kcal/mol]

$$\text{binding energy} = 28.4 \text{ MeV atom}^{-1} \times \frac{23.06 \text{ kcal mole}^{-1}}{1 \text{ eV atom}^{-1}}$$

$$= 655,000,000 \text{ kcal mole}^{-1}$$

Compare this energy with the 83 kcal mole^{-1} required to break carbon-carbon bonds in chemical reactions.)

Every atomic nucleus is lighter than the sum of the masses of the nucleons from which it is built, and this mass loss corresponds to the binding energy of the nucleus. The relative stability of two nuclei with different numbers of nucleons can be assessed by comparing their *mass loss per nucleon*.

The mass loss or binding energy per nuclear particle (protons and neutrons) rises rapidly to a maximum at iron, then falls. Iron is the most stable nucleus of all. The mass losses or binding energies per nucleon are plotted above for all nuclei from helium through uranium.

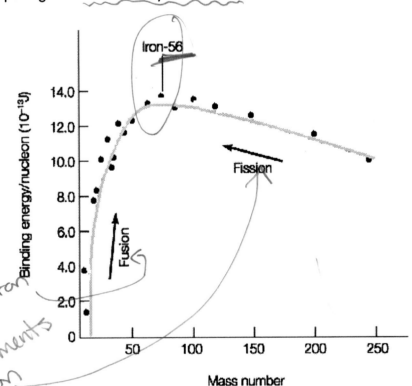

[handwritten: Fusion → fuses elements lighter than Iron]

[handwritten: Fission → divides elements heavier than iron]

After some initial minor irregularities in the first- and second-row elements, the values settle down to a smooth curve, which rises to a maximum at iron, then begins a long descending slope through uranium and beyond.

Iron is the most stable nucleus of all. For elements with smaller atomic numbers than iron, fusion of nuclei to produce heavier elements releases energy, because the products are lighter and more stable on a per-nucleon basis than the reactants. In contrast, beyond iron, fusion absorbs energy because the products are heavier on a per-nucleon basis than the reactants.

Example.
What is the loss of mass per nucleon for the $^{56}_{26}Fe$ atom, compared with its component protons, neutrons, and electrons?

Solution.
The $^{56}_{26}Fe$ atom contains 26 protons, 26 electrons, and 30 neutrons, so the mass calculation is performed as follows::

Mass of 26 protons:	26 (1.0073) amu =	26.1898 amu
Mass of 30 neutrons:	30(1.0087) amu =	30.261 amu
Mass of 26 electrons:	26(0.00055) amu =	0.0143 amu
Total mass of components in iron-56		56.465 amu
Mass of iron-56 atom		55.93 amu
Total mass loss:		0.54 amu
Mass loss/nucleon:	0.54amu/56 =	0.0096 amu

Notice that the loss of mass per nucleon, and hence the binding energy per nucleon, is greater for iron (0.0096 amu) than for helium (0.0076 amu). This means that the iron nucleus is more stable relative to protons and neutrons than the helium nucleus is. If some combination of helium nuclei could be induced to produce an iron nucleus, energy would be given off, which would correspond to the increased stability of the product nucleus per nuclear particle.

COMPETENCY 0028 UNDERSTAND FACTORS AND PROCESSES RELATED TO THE RELEASE OF CHEMICALS INTO THE ENVIRONMENT

A baby born today in the United States is expected to live 30 years longer on an average than a baby born 100 years ago. A significant cause of this improvement is due to chemical technology. The manufacture and distribution of vaccines and antibiotics, an increase in understanding human nutritional needs, and the new drugs that are available now have all played a significant role in improving the length and the quality of human life. But the benefits of these technologies are almost always accompanied by problems and significant risks.

Pesticides are used to control or kill organisms that compete with humans for food, spread disease, or are considered a nuisance. Herbicides are pesticides that attack weeds; insecticides attack insects; fungicides attack molds and other fungus. Sulfur was used as a fungicide in ancient times. The development and use of new pesticides has exploded over the last 60 years, but these pesticides are often poisonous to humans.

The **insecticide DDT** was widely used in the 1940s and 1950s and is responsible for **eradicating malaria from Europe and North America**. It quickly became the most widely used pesticide in the world. In the 1960s, some claimed that DDT was preventing fish-eating birds from reproducing and that it was causing birth defects in humans. DDT is now banned in many countries, but it is still used in developing nations to prevent diseases carried by insects. Unfortunately, its use in agriculture has often led to resistant mosquito strains that have hindered its effectiveness to prevent diseases.

The **herbicide *Roundup*** kills all natural plants it encounters. It began to be used in the 1990s in combination with **genetically engineered crops** that include a gene intended to make the crop (and only the crop) resistant to the herbicide. This combination of chemical and genetic technology has been an economic success but it has raised many concerns about potential problems in the future.

Most scientists believe the emission of greenhouse gases has already led to global warming due to an increase in the greenhouse effect. The greenhouse effect occurs when these gases in the atmosphere warm the planet by absorbing heat to prevent it from escaping into space. This is similar—but not identical—to what occurs in greenhouse buildings. Greenhouse buildings warm an interior space by preventing mixing with colder gases outside.

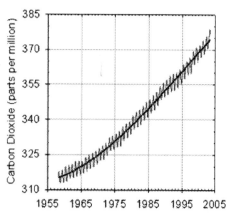

Source: Mauna Loa record, National Oceanic and Atmospheric

CHEMISTRY 238

Many greenhouse gases such as water vapor occur naturally and are important for life to exist on Earth. Human production of carbon dioxide from combustion of fossil fuels has increased the concentration of this important greenhouse gas to its highest value since millions of years ago. The precise impact of these changes in the atmosphere is difficult to predict and is a topic of international concern and political debate.

Rain with a pH less than 5.6 is known as **acid rain**. Acid rain is caused by burning fossil fuels (especially coal) and by fertilizers used in intensive agriculture. These activities emit sulfur and nitrogen in gas compounds that are converted to sulfur oxides and nitrogen oxides. These in turn create sulfuric acid and nitric acid in rain. Acid rain may also be created from gases emitted by volcanoes and other natural sources. Acid rain harms fish and trees and triggers the release of metal ions from minerals into water that can harm people. The problem of acid rain in the United States has been addressed in recent decades by the use of **scrubbers** in coal burning power plants and **catalytic converters** in vehicles.

Ozone is O_3. The ozone layer is a region of the stratosphere that contains higher concentrations of ozone than other parts of the atmosphere. The ozone layer is important for human health because it blocks ultraviolet radiation from the sun, and this helps to protect us from skin cancer. Research in the 1970s revealed that several gases used for refrigeration and other purposes were depleting the ozone layer. Many of these ozone-destroying molecules are short alkyl halides known as chlorofluorocarbons or CFCs. CCl_3F is one example. The widespread use of ozone-destroying gases was banned by an international agreement in the early 1990s. Other substances are used in their place such as CF_3CH_2F, a hydrofluorocarbon. Since that time the concentration of ozone-depleting gases in the atmosphere has been declining and the rate of ozone destruction has been decreasing. Many see this improvement as the most important positive example of

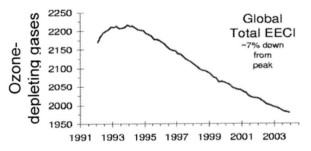

Source: National Oceanic and Atmospheric Administration

international cooperation in helping the environment. The story of these new refrigerants is found at http://www.chemcases.com/fluoro/index.htm.

COMPETENCY 0029   UNDERSTAND THE INTERRELATIONSHIPS AMONG CHEMISTRY, SOCIETY, TECHNOLOGY AND OTHER DISCIPLINES.

Science and society are interconnected. Important discoveries in science influence society and have the potential to alter society. For example, the invention of the printing press caused an upheaval in the society and increased scientific endeavors all at the same time. The printing press made for cheap production of books so that common individuals, not just the wealthy, could access literary works. It made it easier to learn to read and write, increasing the education level of all classes of society. It also made it easier to dream of places, things, and worlds not yet discovered. Scientists could work from exactly the same book at the same time, allowing for fewer errors. The printing press also allowed for the scientific revolution to occur since scientists could communicate their research results with a larger community at faster rates. The printing press changed the way people thought and looked at their world.

At the same time, the needs of society drive the direction of scientific investigation. The fear that the world would be dominated by Hitler and a supply of powerful new weapons lead the United States to begin developing its own atomic bomb. The need to end World War II strengthened and hastened the need for the Manhattan Project to be successful. On August 6, 1945 the world changed forever. The creation of the atomic bomb changed the way society existed. It also changed the direction of science. Would we have discovered nuclear power, the television and computer, nuclear medicine and all of the millions of other discoveries that came from the work of the Manhattan Project scientists? Probably, but the pace and the development of such technologies might not have occurred so quickly.

The union of science, technology, and mathematics has shaped the world we live in today. Science describes the world. It attempts to explain how nature works; from our own bodies to the tiny particles making up the world, from the entire earth to the worlds beyond. Science lets us know in advance what will happen when a cell splits or when two chemicals react. Science is ever-changing. Throughout history, people have developed and validated many different ideas about the flow of the universe. Changing explanations as technology develops allows for new information to emerge.

Technology is the use of scientific knowledge to solve problems. For example, science studies the flow of electrons but technology takes the flow of electrons to create a supercomputer. Mathematics, in turn provides the language to allow this knowledge to be communicated. It creates models for science to use to explain natural phenomena.

The economy is dependent upon the existence of technology. The job market changes as new technologies develop.

For example, with the advent of computer technology, many trained workers are needed, moving our economy from the post-war production line economy to a knowledge-economy based on information technology. This economy is driven by start-ups and entrepreneurs looking for innovation but not radical breakthroughs. It relies on sources of capitol from venture capitalists, IPOs and investors. It is also forcing a new kind of research to occur. Industrial labs are being redefined or eliminated, creating a convergence between scientific disciplines and engineering, providing for new entrepreneur opportunities.

At the same time, advances in scientific knowledge and technology often present ethical dilemmas for society. Industrialization brought the consumption of great amounts of energy, for example. This in turn creates environmental problems, leads to wars, and a depletion of natural resources. Scientific knowledge tells us there is oil is shale that we have not accessed yet. Technology developments will allow us to reach the oil and power our economy. At the same time, new technology will emerge providing alternative power supplies like wind farms or soy fuels. New types of skills will be needed to support these new technologies and the job market; education and our society will shift in response.

The need for technology development drives science, and discoveries in science drive society. Arguably, the first personal computers were mass-marketed by IBM and Apple in the late 1970's. The IBM 5100 had 16K memory, a five inch screen, and cost nearly $11,000 in 1975! A mere twenty-five years later, computers are everywhere and used for everything. They have gigabytes of memory and monitors larger than some televisions! Why did computers develop so quickly? Because of the interconnection between science, technology, and society.

Sample Test

Directions: Read each item and select the best response.

1. A piston compresses a gas at constant temperature. Which gas properties increase?

   I. Average speed of molecules
   II. Pressure
   III. Molecular collisions with container walls per second

   A. I and II
   B. I and III
   C. II and III
   D. I, II, and III

2. The temperature of a liquid is raised at atmospheric pressure. Which liquid property increases?

   A. critical pressure
   B. vapor pressure
   C. surface tension
   D. viscosity

3. Potassium crystallizes with two atoms contained in each unit cell. What is the mass of potassium found in a lattice 1.00×10^6 unit cells wide, 2.00×10^6 unit cells high, and 5.00×10^5 unit cells deep?

   A. 85.0 ng
   B. 32.5 μg
   C. 64.9 μg
   D. 130. μg

4. A gas is heated in a sealed container. Which of the following occur?

   A. gas pressure rises
   B. gas density decreases
   C. the average distance between molecules increases
   D. all of the above

5. How many molecules are in 2.20 pg of a protein with a molecular weight of 150. kDa?

   A. 8.83×10^9
   B. 1.82×10^9
   C. 8.83×10^6
   D. 1.82×10^6

6. At STP, 20. μL of O_2 contain 5.4×10^{16} molecules. According to Avogadro's hypothesis, how many molecules are in 20. μL of Ne?

   A. 5.4×10^{15}
   B. 1.0×10^{16}
   C. 2.7×10^{16}
   D. 5.4×10^{16}

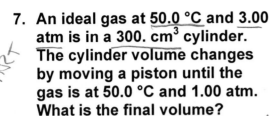

7. An ideal gas at 50.0 °C and 3.00 atm is in a 300. cm³ cylinder. The cylinder volume changes by moving a piston until the gas is at 50.0 °C and 1.00 atm. What is the final volume?

   A. 100. cm³
   B. 450. cm³
   C. 900. cm³
   D. 1.20 dm³

8. Which gas law may be used to solve the previous question?

   A. Charles's law
   B. Boyle's law
   C. Graham's law
   D. Avogadro's law

9. A blimp is filled with 5000. m³ of helium at 28.0 °C and 99.7 kPa. What is the mass of helium used?

   $R = 8.3144 \dfrac{J}{mol\text{-}K}$

   A. 797 kg
   B. 810. kg
   C. 879 kg
   D. 8.57×10^3 kg

10. Which of the following are able to flow from one place to another?

    I.   Gases
    II. Liquids
    III. Solids
    IV. Supercritical fluids

    A. I and II
    B. II only
    C. I, II, and IV
    D. I, II, III, and IV

11. One mole of an ideal gas at STP occupies 22.4 L. At what temperature will one mole of an ideal gas at one atm occupy 31.0 L?

    A. 34.6 °C
    B. 105 °C
    C. 378 °C
    D. 442 °C

12. Why does $CaCl_2$ have a higher normal melting point than NH_3?

    A. Covalent bonds are stronger than London dispersion forces.
    B. Covalent bonds are stronger than hydrogen bonds.
    C. Ionic bonds are stronger than London dispersion forces.
    D. Ionic bonds are stronger than hydrogen bonds.

13. Which intermolecular attraction explains the following trend in straight-chain alkanes?

Condensed structural formula	Boiling point (°C)
CH_4	-161.5
CH_3CH_3	-88.6
$CH_3CH_2CH_3$	-42.1
$CH_3CH_2CH_2CH_3$	-0.5
$CH_3CH_2CH_2CH_2CH_3$	36.0
$CH_3CH_2CH_2CH_2CH_2CH_3$	68.7

A. London dispersion forces
B. Dipole-dipole interactions
C. Hydrogen bonding
D. Ion-induced dipole interactions

14. List the substances NH_3, PH_3, $MgCl_2$, Ne, and N_2 in order of increasing melting point.

A. N_2 < Ne < PH_3 < NH_3 < $MgCl_2$
B. N_2 < NH_3 < Ne < $MgCl_2$ < PH_3
C. Ne < N_2 < NH_3 < PH_3 < $MgCl_2$
D. Ne < N_2 < PH_3 < NH_3 < $MgCl_2$

15. 1-butanol, ethanol, methanol, and 1-propanol are all liquids at room temperature. Rank them in order of increasing viscosity.

A. 1-butanol < 1-propanol < ethanol < methanol
B. methanol < ethanol < 1-propanol < 1-butanol
C. methanol < ethanol < 1-butanol < 1-propanol
D. 1-propanol < 1-butanol < ethanol < methanol

16. Which gas has a diffusion rate of 25% the rate for hydrogen?

A. helium
B. methane
C. nitrogen
D. oxygen

17. 2.00 L of an unknown gas at 1500. mm Hg and a temperature of 25.0 °C weighs 7.52 g. Assuming the ideal gas equation, what is the molecular mass of the gas?

760 mm Hg=1 atm
R=0.08206 L-atm/(mol-K)

A. 21.6 u
B. 23.3 u
C. 46.6 u
D. 93.2 u

18. Which substance is most likely to be a gas at room temperature?

A. SeO_2
B. F_2
C. $CaCl_2$
D. I_2

19. What pressure is exerted by a mixture of 2.7 g of H_2 and 59 g of Xe at STP on a 50. L container?

A. 0.69 atm
B. 0.76 atm
C. 0.80 atm
D. 0.97 atm

20. A few minutes after opening a bottle of perfume, the scent is detected on the other side of the room. What law relates to this phenomenon?

A. Graham's law
B. Dalton's law
C. Boyle's law
D. Avogadro's law

21. Which of the following are true?

A. Solids have no vapor pressure.
B. Dissolving a solute in a liquid increases its vapor pressure.
C. The vapor pressure of a pure substance is characteristic of that substance and its temperature.
D. All of the above

22. Find the partial pressure of N_2 in a container at 150. kPa holding H_2O and N_2 at 50 °C. The vapor pressure of H_2O at 50 °C is 12 kPa.

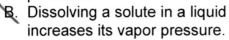

A. 12 kPa
B. 138 kPa
C. 162 kPa
D. The value cannot be determined.

23. The normal boiling point of water on the Kelvin scale is closest to:

A. 112 K
B. 212 K
C. 273 K
D. 373 K
E.

24. Which phase may be present at the triple point of a substance?

I. Gas
II. Liquid
III. Solid
IV. Supercritical fluid

A. I, II, and III
B. I, II, and IV
C. II, III, and IV
D. I, II, III, and IV

25. In the following phase diagram, _____ occurs as P is decreased from A to B at constant T and _____ occurs as T is increased from C to D at constant P.

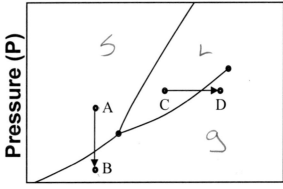

Temperature (T)

A. deposition, melting
B. sublimation, melting
C. deposition, vaporization
D. sublimation, vaporization

26. Heat is added to a pure solid at its melting point until it all becomes liquid at its freezing point. Which of the following occur?

A. Intermolecular attractions are weakened.
B. The kinetic energy of the molecules does not change.
C. The freedom of the molecules to move about increases.
D. All of the above.

27. Which of the following occur when NaCl dissolves in water?

A. Heat is required to break bonds in the NaCl crystal lattice.
B. Heat is released when hydrogen bonds in water are broken.
C. Heat is required to form bonds of hydration.
D. The oxygen end of the water molecule is attracted to the Cl^- ion.

28. The solubility of $CoCl_2$ is 54 g per 100 g of ethanol. Three flasks each contain 100 g of ethanol. Flask #1 also contains 40 g $CoCl_2$ in solution. Flask #2 contains 56 g $CoCl_2$ in solution. Flask #3 contains 5 g of solid $CoCl_2$ in equilibrium with 54 g $CoCl_2$ in solution. Which of the following describe the solutions present in the liquid phase of the flasks?

A. #1-saturated, #2-supersaturated, #3-unsaturated.
B. #1-unsaturated, #2-miscible, #3-saturated.
C. #1-unsaturated, #2-supersaturated, #3-saturated.
D. #1-unsaturated, #2-not at equilibrium, #3-miscible.

29. The solubility at 1.0 atm of pure CO_2 in water at 25 °C is 0.034 M. According to Henry's law, what is the solubility at 4.0 atm of pure CO_2 in water at 25 °C? Assume no chemical reaction occurs between CO_2 and H_2O.

A. 0.0085 M
B. 0.034 M
C. 0.14 M
D. 0.25 M

30. Carbonated water is bottled at 25 °C under pure CO_2 at 4.0 atm. Later the bottle is opened at 4 °C under air at 1.0 atm that has a partial pressure of 3×10^{-4} atm CO_2. Why do CO_2 bubbles form when the bottle is opened?

   A. CO_2 falls out of solution due to a drop in solubility at the lower total pressure.
   B. CO_2 falls out of solution due to a drop in solubility at the lower CO_2 pressure.
   C. CO_2 falls out of solution due to a drop in solubility at the lower temperature.
   D. CO_2 is formed by the decomposition of carbonic acid.

31. When KNO₃ dissolves in water, the water grows slightly colder. An increase in temperature will _____ the solubility of KNO_3.

   A. increase
   B. decrease
   C. have no effect on
   D. have an unknown effect with the information given on

32. An experiment requires 100. mL of a 0.500 M solution of $MgBr_2$. How many grams of $MgBr_2$ will be present in this solution?

   A. 9.21 g
   B. 11.7 g
   C. 12.4 g
   D. 15.6 g

33. 500. mg of RbOH are added to 500. g of ethanol (C_2H_6O) resulting in 395 mL of solution. Determine the molarity and molality of RbOH.

   A. 0.0124 M, 0.00488 m
   B. 0.0124 M, 0.00976 m
   C. 0.0223 M, 0.00488 m
   D. 0.0223 M, 0. 00976 m

34. 20.0 g H_3PO_4 in 1.5 L of solution are intended to react with KOH according to the following reaction:
   $$H_3PO_4 + 3\,KOH \rightarrow K_3PO_4 + 3\,H_2O$$
   What is the molarity and normality of the H_3PO_4 solution?

   A. 0.41 M, 1.22 N
   B. 0.41 M, 0.20 N
   C. 0.14 M, 0.045 N
   D. 0.14 M, 0. 41 N

35. Aluminum sulfate is a strong electrolyte. What is the concentration of all species in a 0.2 M solution of aluminum sulfate?

   A. 0.2 M Al^{3+}, 0.2 M SO_4^{2-}
   B. 0.4 M Al^{3+}, 0.6 M SO_4^{2-}
   C. 0.6 M Al^{3+}, 0.4 M SO_4^{2-}
   D. 0.2 M $Al_2(SO_4)_3$

36. 15 g of formaldehyde (CH_2O) are dissolved in 100. g of water. Calculate the weight percentage and mole fraction of formaldehyde in the solution.

    A. 13%, 0.090
    B. 15%, 0.090
    C. 13%, 0.083
    D. 15%, 0.083

37. Which of the following would make the best solvent for Br_2?

    A. H_2O
    B. CS_2
    C. NH_3
    D. molten NaCl

38. Which of the following is most likely to dissolve in water?

    A. H_2
    B. CCl_4
    C. SF_6
    D. CH_3OH

39. Which of the following is not a colligative property?

    A. Viscosity lowering
    B. Freezing point lowering
    C. Boiling point elevation
    D. Vapor pressure lowering

40. $$BaCl_2(aq) + Na_2SO_4(aq) \rightarrow$$
    $$BaSO_4(s) + 2NaCl(aq)$$
    is an example of a _____ reaction.

    A. acid-base
    B. precipitation
    C. redox
    D. nuclear

41. List the following aqueous solutions in order of increasing boiling point.

    I.   0.050 m $AlCl_3$
    II. 0.080 m $Ba(NO_3)_2$
    III. 0.090 m NaCl
    IV. 0.12 m ethylene glycol ($C_2H_6O_2$)

    A. I < II < III < IV
    B. I < III < IV < II
    C. IV < III < I < II
    D. IV < III < II < I

42. Osmotic pressure is the pressure required to prevent _____ flowing from low to high _____ concentration across a semipermeable membrane.

    A. solute, solute
    B. solute, solvent
    C. solvent, solute
    D. solvent, solvent

43. A solution of NaCl in water is heated on a mountain in an open container until it boils at 100. °C. The air pressure on the mountain is 0.92 atm. According to Raoult's law, what mole fraction of Na^+ and Cl^- are present in the solution?

    A. 0.04 Na^+, 0.04 Cl^-
    B. 0.08 Na^+, 0.08 Cl^-
    C. 0.46 Na^+, 0.46 Cl^-
    D. 0.92 Na^+, 0.92 Cl^-

44. Write a balanced nuclear equation for the emission of an alpha particle by polonium-209.

    A. $^{209}_{84}Po \rightarrow ^{205}_{81}Pb + ^4_2He$
    B. $^{209}_{84}Po \rightarrow ^{205}_{82}Bi + ^4_2He$
    C. $^{209}_{84}Po \rightarrow ^{209}_{85}At + ^0_{-1}e$
    D. $^{209}_{84}Po \rightarrow ^{205}_{82}Pb + ^4_2He$

45. Write a balanced nuclear equation for the decay of calcium-45 to scandium-45.

    A. $^{45}_{20}Ca \rightarrow ^{41}_{18}Sc + ^4_2He$
    B. $^{45}_{20}Ca + ^0_1e \rightarrow ^{45}_{21}Sc$
    C. $^{45}_{20}Ca \rightarrow ^{45}_{21}Sc + ^0_{-1}e$
    D. $^{45}_{20}Ca + ^0_1p \rightarrow ^{45}_{21}Sc$

46. ^3_1H decays with a half-life of 12 years. 3.0 g of pure ^3_1H were placed in a sealed container 24 years ago. How many grams of ^3_1H remain?

    A. 0.38 g
    B. 0.75 g
    C. 1.5 g
    D. 3.0 g

47. Oxygen-15 has a half-life of 122 seconds. What percentage of a sample of oxygen-15 has decayed after 300. seconds?

    A. 18.2%
    B. 21.3%
    C. 78.7%
    D. 81.8%

48. Which of the following isotopes is commonly used for medical imaging in the diagnose of diseases?

    A. cobalt-60
    B. technetium-99m
    C. tin-117m
    D. plutonium-238

49. Carbon-14 dating would be useful in obtaining the age of which object?

    A. a 20th century Picasso painting
    B. a mummy from ancient Egypt
    C. a dinosaur fossil
    D. all of the above

50. Which of the following isotopes can create a chain reaction of nuclear fission?

A. uranium-235
B. uranium-238
C. plutonium-238
D. all of the above

51. List the following scientists in chronological order from earliest to most recent with respect to their most significant contribution to atomic theory:

I. John Dalton
II. Niels Bohr
III. J. J. Thomson
IV. Ernest Rutherford

A. I, III, II, IV
B. I, III, IV, II
C. I, IV, III, II
D. III, I, II, IV

52. Match the theory with the scientist who first proposed it:

I. Electrons, atoms, and all objects with momentum also exist as waves.
II. Electron density may be accurately described by a single mathematical equation.
III. There is an inherent indeterminacy in the position and momentum of particles.
IV. Radiant energy is transferred between particles in exact multiples of a discrete unit.

A. I-de Broglie, II-Planck, III-Schrödinger, IV-Thomson
B. I-Dalton, II-Bohr, III-Planck, IV-de Broglie
C. I-Henry, II-Bohr, III-Heisenberg, IV-Schrödinger
D. I-de Broglie, II-Schrödinger, III-Heisenberg, IV-Planck

53. How many neutrons are in $^{60}_{27}Co$?

A. 27
B. 33
C. 60
D. 87

54. The terrestrial composition of an element is: 50.7% as an isotope with an atomic mass of 78.9 u and 49.3% as an isotope with an atomic mass of 80.9 u. Both isotopes are stable. Calculate the atomic mass of the element.
   A. 79.0 u
   B. 79.8 u
   C. 79.9 u
   D. 80.8 u

55. Which of the following is a correct electron arrangement for oxygen?

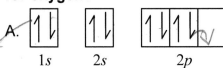

   A.

   B. $1s^2 1p^2 2s^2 2p^2$
   C. 2, 2, 4
   D. none of the above

56. Which of the following statements about radiant energy is **not** true?

   A. The energy change of an electron transition is directly proportional to the wavelength of the emitted or absorbed photon.
   B. The energy of an electron in a hydrogen atom depends only on the principle quantum number.
   C. The frequency of photons striking a metal determines whether the photoelectric effect will occur.
   D. The frequency of a wave of electromagnetic radiation is inversely proportional to its wavelength

57. Match the orbital diagram for the ground state of carbon with the rule/principle it violates:

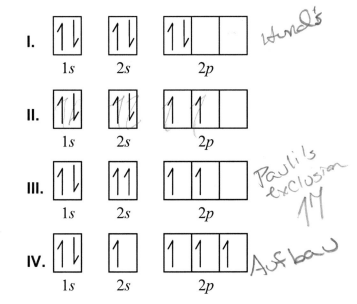

   A. I-Pauli exclusion, II-Aufbau, III-no violation, IV-Hund's
   B. I-Aufbau, II-Pauli exclusion, III-no violation, IV-Hund's
   C. I-Hund's, II-no violation, III-Pauli exclusion, IV-Aufbau
   D. I-Hund's, II-no violation, III-Aufbau, IV-Pauli exclusion

58. Select the list of atoms that are arranged in order of increasing size.

   A. Mg, Na, Si, Cl
   B. Si, Cl, Mg, Na
   C. Cl, Si, Mg, Na
   D. Na, Mg, Si, Cl

59. Based on trends in the periodic table, which of the following properties would you expect to be greater for Rb than for K?

 I.   Density
 II. Melting point
 III. Ionization energy
 IV. Oxidation number in a compound with chlorine

 A. I only
 B. I, II, and III
 C. II and III
 D. I, II, III, and IV

60. Which oxide forms the strongest acid in water?

 A. Al_2O_3
 B. Cl_2O_7
 C. As_2O_5
 D. CO_2

61. Rank the following bonds from least to most polar:

 C-H, C-Cl, H-H, C-F

 A. C-H < H-H < C-F < C-Cl
 B. H-H < C-H < C-F < C-Cl
 C. C-F < C-Cl < C-H < H-H
 D. H-H < C-H < C-Cl < C-F

62. At room temperature, $CaBr_2$ is expected to be:

 A. a ductile solid
 B. a brittle solid
 C. a soft solid
 D. a gas

63. Which of the following is a proper Lewis dot structure of CHClO?

A.

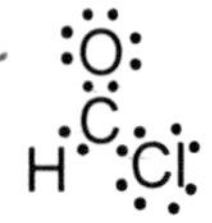

B.

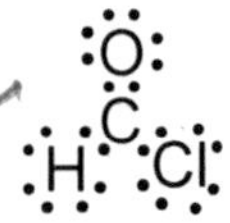

C.

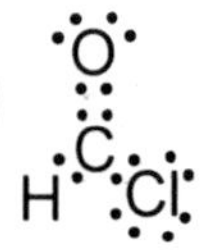

D.

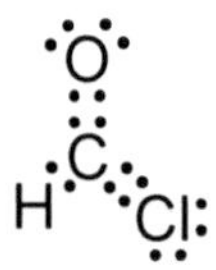

64. In C_2H_2, each carbon atom contains the following valence orbitals:

 A. p only
 B. p and sp hybrids
 C. p and sp^2 hybrids
 D. sp^3 hybrids only

65. Which statement about molecular structures is false?

A. 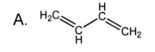 is a conjugated molecule

B. A bonding σ orbital connects two atoms by the straight line between them.

C. A bonding π orbital connects two atoms in a separate region from the straight line between them.

D. The anion with resonance forms

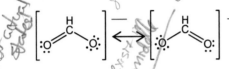

will always exist in one form or the other.

66. What is the shape of the PH_3 molecule? Use the VSEPR model.

A. Trigonal pyramidal
B. Trigonal bipyramidal
C. Trigonal planar
D. Tetrahedral

67. What is the chemical composition of magnesium nitrate?

A. 11.1% Mg, 22.2% N, 66.7% O
B. 16.4% Mg, 18.9% N, 64.7% O
C. 20.9% Mg, 24.1% N, 55.0% O
D. 28.2% Mg, 16.2% N, 55.7% O

68. The IUPAC name for Cu_2SO_3 is:

A. Dicopper sulfur trioxide
B. Copper (II) sulfate
C. Copper (I) sulfite
D. Copper (II) sulfite

69. Which name or formula is not represented properly?

A. Cl_4S
B. $KClO_3$
C. Calcium dihydrogen phosphate
D. Sulfurous acid

70. Household "chlorine bleach" is sodium hypochlorite. Which of the following best represent the production of sodium hypochlorite, sodium chloride, and water by bubbling chlorine gas through aqueous sodium hydroxide?

A. $4Cl(g) + 4NaOH(aq) \rightarrow$
$NaClO_2(aq) + 3NaCl(aq) + 2H_2O(l)$

B. $2Cl_2(g) + 4NaOH(aq) \rightarrow$
$NaClO_2(aq) + 3NaCl(aq) + 2H_2O(l)$

C. $2Cl(g) + 2NaOH(aq) \rightarrow$
$NaClO(aq) + NaCl(aq) + H_2O(l)$

D. $Cl_2(g) + 2NaOH(aq) \rightarrow$
$NaClO(aq) + NaCl(aq) + H_2O(l)$

71. Balance the equation for the neutralization reaction between phosphoric acid and calcium hydroxide by filling in the blank stoichiometric coefficients.

$$\underline{2}\,H_3PO_4 + \underline{3}\,Ca(OH)_2 \rightarrow$$
$$\underline{}\,Ca_3(PO_4)_2 + \underline{6}\,H_2O$$

- A. 4, 3, 1, 4
- B. 2, 3, 1, 8
- C. 2, 3, 1, 6
- D. 2, 1, 1, 2

72. Write an equation showing the reaction between calcium nitrate and lithium sulfate in aqueous solution. Include all products.

A. $CaNO_3(aq) + Li_2SO_4(aq) \rightarrow$
$CaSO_4(s) + Li_2NO_3(aq)$

B. $Ca(NO_3)_2(aq) + Li_2SO_4(aq) \rightarrow$
$CaSO_4(s) + 2LiNO_3(aq)$

C. $Ca(NO_3)_2(aq) + Li_2SO_4(aq) \rightarrow$
$2LiNO_3(s) + CaSO_4(aq)$

D. $Ca(NO_3)_2(aq) + Li_2SO_4(aq) + 2H_2O(l) \rightarrow$
$2LiNO_3(aq) + Ca(OH)_2(aq) + H_2SO_4(aq)$

73. Find the mass of CO_2 produced by the combustion of 15 kg of isopropyl alcohol in the reaction:

$$2C_3H_7OH + 9O_2 \rightarrow 6CO_2 + 8H_2O$$

- A. 33 kg
- B. 44 kg
- C. 50 kg
- D. 60 kg

74. What is the density of nitrogen gas at STP? Assume an ideal gas and a value of 0.08206 L·atm/(mol·K) for the gas constant.

- A. 0.62 g/L
- B. 1.14 g/L
- C. 1.25 g/L
- D. 2.03 g/L

75. Find the volume of methane that will produce 12 m³ of hydrogen in the reaction:

$$CH_4(g) + H_2O(g) \rightarrow CO(g) + 3H_2(g)$$

Assume temperature and pressure remain constant.

- A. 4.0 m³
- B. 32 m³
- C. 36 m³
- D. 64 m³

76. A 100. L vessel of pure O_2 at 500. kPa and 20. °C is used for the combustion of butane:

$2C_4H_{10} + 13O_2 \rightarrow 8CO_2 + 10H_2O$

Find the mass of butane to consume all the O_2 in the vessel. Assume O_2 is an ideal gas and use a value of $R = 8.314$ J/(mol·K).

A. 183 g
B. 467 g
C. 1.83 kg
D. 7.75 kg

77. Consider the reaction between iron and hydrogen chloride gas:

$Fe(s) + 2HCl(g) \rightarrow FeCl_2(s) + H_2(g)$

7 moles of iron and 10 moles of HCl react until the limiting reagent is consumed. Which statements are true?

I.   HCl is the excess reagent
II. HCl is the limiting reagent
III. 7 moles of H_2 are produced
IV. 2 moles of the excess reagent remain

A. I and III
B. I and IV
C. II and III
D. II and IV

78. 32.0 g of hydrogen and 32.0 grams of oxygen react to form water until the limiting reagent is consumed. What is present in the vessel after the reaction is complete?

A. 16.0 g O_2 and 48.0 g H_2O
B. 24.0 g H_2 and 40.0 g H_2O
C. 28.0 g H_2 and 36.0 g H_2O
D. 28.0 g H_2 and 34.0 g H_2O

79. Three experiments were performed at the same initial temperature and pressure to determine the rate of the reaction

$2ClO_2(g) + F_2(g) \rightarrow 2ClO_2F(g)$

Results are shown in the table below. Concentrations are given in millimoles per liter (mM).

Exp.	Initial $[ClO_2]$ (mM)	Initial $[F_2]$ (mM)	Initial rate of $[ClO_2F]$ increase (mM/sec)
1	5.0	5.0	0.63
2	5.0	20	2.5
3	10	10	2.5

What is the rate law for this reaction?

A. Rate $= k[F_2]$

B. Rate $= k[ClO_2][F_2]$

C. Rate $= k[ClO_2]^2[F_2]$

D. Rate $= k[ClO_2][F_2]^2$

80. The reaction

$$(CH_3)_3CBr(aq) + OH^-(aq) \rightarrow$$

$$(CH_3)_3COH(aq) + Br^-(aq)$$

occurs in three elementary steps:

$(CH_3)_3CBr \rightarrow (CH_3)_3C^+ + Br^-$ is slow

$(CH_3)_3C^+ + H_2O \rightarrow (CH_3)_3COH_2^+$ is fast

$(CH_3)_3COH_2^+ + OH^- \rightarrow$
$(CH_3)_3COH + H_2O$ is fast

What is the rate law for this reaction?

A. $Rate = k\left[(CH_3)_3CBr\right]$

B. $Rate = k\left[OH^-\right]$

C. $Rate = k\left[(CH_3)_3CBr\right]\left[OH^-\right]$

D. $Rate = k\left[(CH_3)_3CBr\right]^2$

81. **Which statement about equilibrium is not true?**

A. Equilibrium shifts to minimize the impact of changes.

B. Forward and reverse reactions have equal rates at equilibrium.

C. A closed container of air and water is at a vapor-liquid equilibrium if the humidity is constant.

D. The equilibrium between solid and dissolved forms is maintained when salt is added to an unsaturated solution.

82. **Which statements about reaction rates are true?**

I. Catalysts shift an equilibrium to favor product formation.

II. Catalysts increase the rate of forward and reverse reactions.

III. A greater temperature increases the chance that a molecular collision will overcome a reaction's activation energy.

IV. A catalytic converter contains a homogeneous catalyst.   *heterogeneous*

A. I and II

B. II and III

C. II, III, and IV

D. I, III, and IV

83. **Write the equilibrium expression K_{eq} for the reaction**

$$CO_2(g) + H_2(g) \rightleftharpoons CO(g) + H_2O(l)$$

A. $\dfrac{[CO][H_2O]}{[CO_2][H_2]^2}$

B. $\dfrac{[CO_2][H_2]}{[CO][H_2O]}$

C. $\dfrac{[CO][H_2O]}{[CO_2][H_2]}$

D. $\dfrac{[CO]}{[CO_2][H_2]}$

84. What could cause this change in the energy diagram of a reaction?

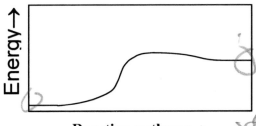

Reaction pathway→

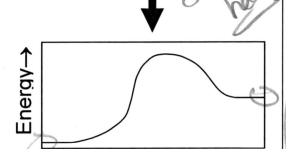

Reaction pathway→

A. Adding catalyst to an endothermic reaction
B. Removing catalyst from an endothermic reaction
C. Adding catalyst to an exothermic reaction
D. Removing catalyst from an exothermic reaction

85. $BaSO_4$ ($K_{sp} = 1 \times 10^{-10}$) is added to pure H_2O. How much is dissolved in 1 L of saturated solution?

A. 2 mg
B. 10 μg
C. 2 μg
D. 100 pg

86. The exothermic reaction
$2NO(g) + Br_2(g) \rightleftharpoons 2NOBr(g)$ is at equilibrium. According to LeChatelier's principle:

A. Adding Br_2 will increase [NO].
B. An increase in container volume (with T constant) will increase [NOBr].
C. An increase in pressure (with T constant) will increase [NOBr].
D. An increase in temperature (with P constant) will increase [NOBr].

87. At a certain temperature, T, the equilibrium constant for the reaction
$2NO(g) \rightleftharpoons N_2(g) + O_2(g)$ is
$K_{eq} = 2 \times 10^3$. If a 1.0 L container at this temperature contains 90 mM N_2, 20 mM O_2, and 5 mM NO, what will occur?

A. The reaction will make more N_2 and O_2.
B. The reaction is at equilibrium.
C. The reaction will make more NO.
D. The temperature, T, is required to solve this problem.

88. Which statement about acids and bases is not true?

A. All strong acids ionize in water.
B. All Lewis acids accept an electron pair.
C. All Brønsted bases use OH^- as a proton acceptor.
D. All Arrhenius acids form H^+ ions in water.

89. Which of the following are listed from weakest to strongest acid?

A. H_2SO_3, H_2SeO_3, H_2TeO_3
B. HBrO, $HBrO_2$, $HBrO_3$, $HBrO_4$
C. HI, HBr, HCl, HF
D. H_3PO_4, $H_2PO_4^-$, HPO_4^{2-}

90. NH_4F is dissolved in water. Which of the following are conjugate acid/base pairs present in the solution?

I. NH_4^+/NH_4OH
II. HF/F^-
III. H_3O^+/H_2O
IV. H_2O/OH^-

A. I, II, and III
B. I, III, and IV
C. II and IV
D. II, III, and IV

91. What are the pH and the pOH of 0.010 M $HNO_3(aq)$?

A. pH = 1.0, pOH = 9.0
B. pH = 2.0, pOH = 12.0
C. pH = 2.0, pOH = 8.0
D. pH = 8.0, pOH = 6.0

92. What is the pH of a buffer made of 0.128 M sodium formate (HCOONa) and 0.072 M formic acid (HCOOH)? The pK_a of formic acid is 3.75.

A. 2.0
B. 3.0
C. 4.0
D. 5.0

93. A sample of 50.0 ml KOH is titrated with 0.100 M $HClO_4$. The initial buret reading is 1.6 ml and the reading at the endpoint is 22.4 ml. What is [KOH]?

A. 0.0416 M
B. 0.0481 M
C. 0.0832 M
D. 0.0962 mM

94. Rank the following from lowest to highest pH. Assume a small volume for the component given in moles:

I. 0.01 mol HCl added to 1 L H_2O
II. 0.01 mol HI added to 1 L of an acetic acid/sodium acetate solution at pH 4.0
III. 0.01 mol NH_3 added to 1 L H_2O
IV. 0.1 mol HNO_3 added to 1 L of a 0.1 M $Ca(OH)_2$ solution

A. I < II < III < IV
B. I < II < IV < III
C. II < I < III < IV
D. II < I < IV < III

95. The curve below resulted from the titration of a _____ _____ with a _____ _____ titrant.

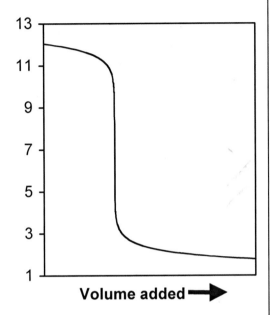

Volume added ➡

A. weak acid, strong base
B. weak base, strong acid
C. strong acid, strong base
D. strong base, strong acid

96. Which statement about thermochemistry is true?

A. Particles in a system move about less freely at high entropy
B. Water at 100 °C has the same internal energy as water vapor at 100°C
C. A decrease in the order of a system corresponds to an increase in entropy. ✓
D. At its sublimation temperature, dry ice has a higher entropy than gaseous CO_2

97. What is the heat change of 36.0 g H_2O at atmospheric pressure when its temperature is reduced from 125 °C to 40. °C? Use the following data:

Values for water	
Heat capacity of solid	37.6 J/mol•°C
Heat capacity of liquid	75.3 J/mol•°C
Heat capacity of gas	33.1 J/mol•°C
Heat of fusion	6.02 kJ/mol
Heat of vaporization	40.67 kJ/mol

A. −92.0 kJ
B. −10.8 kJ
C. 10.8 kJ
D. 92.0 kJ

98. What is the standard heat of combustion of $CH_4(g)$? Use the following data:

Standard heats of formation	
$CH_4(g)$	−74.8 kJ/mol
$CO_2(g)$	−393.5 kJ/mol
$H_2O(l)$	−285.8 kJ/mol

A. −890.3 kJ/mol
B. −604.6 kJ/mol
C. −252.9 kJ/mol
D. −182.5 kJ/mol

99. Which reaction creates products at a lower total entropy than the reactants?

A. Dissolution of table salt:
$NaCl(s) \rightarrow Na^+(aq) + Cl^S(aq)$

B. Oxidation of iron:
$4Fe(s) + 3O_2(g) \rightarrow 2Fe_2O_3(s)$

C. Dissociation of ozone:
$O_3(g) \rightarrow O_2(g) + O(g)$

D. Vaporization of butane:
$C_4H_{10}(l) \rightarrow C_4H_{10}(g)$

100. Which statement about reactions is true?

A. All spontaneous reactions are both exothermic and cause an increase in entropy.

B. An endothermic reaction that increases the order of the system cannot be spontaneous.

C. A reaction can be non-spontaneous in one direction and also non-spontaneous in the opposite direction.

D. Melting snow is an exothermic process. requires heat

101. 10. kJ of heat are added to one kilogram of Iron at 10. °C. What is its final temperature? The specific heat of iron is 0.45 J/g•°C.

A. 22 °C
B. 27 °C
C. 32 °C
D. 37 °C

102. Which reaction is <u>not</u> a redox process?

A. Combustion of octane:
$2C_8H_{18} + 25O_2 \rightarrow 16CO_2 + 18H_2O$

B. Depletion of a lithium battery:
$Li + MnO_2 \rightarrow LiMnO_2$

C. Corrosion of aluminum by acid:
$2Al + 6HCl \rightarrow 2AlCl_3 + 3H_2$

D. Taking an antacid for heartburn:
$CaCO_3 + 2HCl \rightarrow CaCl_2 + H_2CO_3$
$\rightarrow CaCl_2 + CO_2 + H_2O$

103. Given the following heats of reaction:
$\Delta H = -0.3$ kJ/mol for
$Fe(s) + CO_2(g) \rightarrow FeO(s) + CO(g)$
$\Delta H = 5.7$ kJ/mol for
$2Fe(s) + 3CO_2(g) \rightarrow Fe_2O_3(s) + 3CO(g)$
and $\Delta H = 4.5$ kJ/mol for
$3FeO(s) + CO_2(g) \rightarrow Fe_3O_4(s) + CO(g)$

use Hess's Law to determine the heat of reaction for:

$3Fe_2O_3(s) + CO(g) \rightarrow 2Fe_3O_4(s) + CO_2(g)$

A. –10.8 kJ/mol
B. –9.9 kJ/mol
C. –9.0 kJ/mol
D. –8.1 kJ/mol

104. What is the oxidant in the reaction:

$$2H_2S + SO_2 \rightarrow 3S + 2H_2O \ ?$$

A. H_2S
B. SO_2
C. S
D. H_2O

105. Molten NaCl is subjected to electrolysis. What reaction takes place at the cathode?

A. $2Cl^-(l) \rightarrow Cl_2(g) + 2e^-$
B. $Cl_2(g) + 2e^- \rightarrow 2Cl^-(l)$
C. $Na^+(l) + e^- \rightarrow Na(l)$
D. $Na^+(l) \rightarrow Na(l) + e^-$

106. What is the purpose of the salt bridge in an electrochemical cell?

A. To receive electrons from the oxidation half-reaction
B. To relieve the buildup of positive charge in the anode half-cell
C. To conduct electron flow
D. To permit positive ions to flow from the cathode half-cell to the anode half-cell

107. Given:
$E°=-2.37V$ for $Mg^{2+}(aq)+2e^- \rightarrow Mg(s)$
and
$E°=0.80$ V for $Ag^+(aq)+e^- \rightarrow Ag(s)$,
what is the standard potential of a voltaic cell composed of a piece of magnesium dipped in a 1 M Ag^+ solution and a piece of silver dipped in 1 M Mg^{2+}?

A. 0.77 V
B. 1.57 V
C. 3.17 V
D. 3.97 V

108.

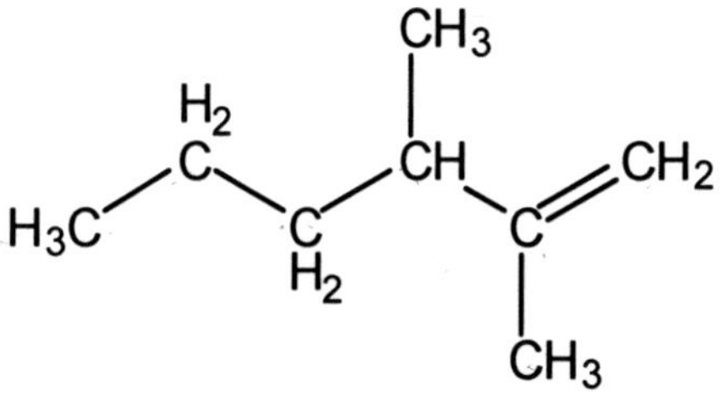

A proper name for this hydrocarbon is:

A. 4,5-dimethyl-6-hexene
B. 2,3-dimethyl-1-hexene
C. 4,5-dimethyl-6-hexyne
D. 2-methyl-3-propyl-1-butene

109. An IUPAC approved name for this molecule is:

   A. butanal
   B. propanal
   C. butanoic acid
   D. propanoic acid

110. Which molecule has a systematic name of methyl ethanoate?

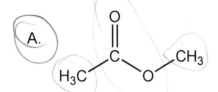

A.

B.

C.

D.

111. This compound

contains an:

   A. alkene, carboxylic acid, ester, and ketone
   B. aldehyde, alkyne, ester, and ketone
   C. aldehyde, alkene, carboxylic acid, and ester
   D. acid anhydride, aldehyde, alkene, and amine

112. Which group of scientists made contributions in the same area of chemistry?

   A. Volta, Kekulé, Faraday, London
   B. Hess, Joule, Kelvin, Gibbs
   C. Boyle, Charles, Arrhenius, Pauli
   D. Davy, Mendeleev, Ramsay, Galvani

113. Which of the following pairs are isomers?

I.

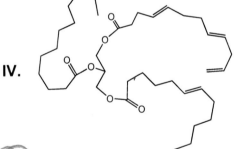

II.   pentanal 2-pentanone

III.

IV.

A. I and IV
B. II and III
C. I, II, and III
D. I, II, III, and IV

114. Which instrument would be most useful for separating two different proteins from a mixture?

A. UV/Vis spectrophotometer
B. Mass spectrometer
C. Gas chromatograph
D. Liquid chromatograph

115. Classify these biochemicals.

I.

II.

III.

IV.

A. I-nucleotide, II-sugar, III-peptide, IV-fat
B. I-disaccharide, II-sugar, III-fatty acid, IV-polypeptide
C. I-disaccharide, II-amino acid, III-fatty acid, IV-polysaccharide
D. I I-nucleotide, II-sugar, III-triacylglyceride, and IV-DNA

116. You create a solution of 2.00 µg/ml of a pigment and divide the solution into 12 samples. You give four samples each to three teams of students. They use a spectrophotometer to determine the pigment concentration. Here is their data:

| Team | Concentration (µg/ml) | | | |
	sample 1	sample 2	sample 3	sample 4
1	1.98	1.93	1.92	1.88
2	1.70	1.72	1.69	1.70
3	1.78	1.99	2.87	2.20

Which of the following are true?

A. Team 1 has the most precise data
B. Team 3 has the most accurate data in spite of it having low precision
C. The data from team 2 is characteristic of a systematic error
D. The data from team 1 is more characteristic of random error than the data from team 3.

117. Which pair of measurements have an identical meaning?

A. 32 micrograms and 0.032 g
B. 26 nm and 2.60×10^{-8} m
C. 3.01×10^{-5} m^3 and 30.1 ml
D. 0.0020 L and 20 cm^3

118. Match the instrument with the quantity it measures

II. eudiometer
III. calorimeter
IV. manometer
V. hygrometer

A. I-volume, II-mass, III-radioactivity, IV-humidity
B. I-volume, II-heat, III-pressure, IV-humidity
C. I-viscosity, II-mass, III-pressure, IV-surface tension
D. I-viscosity, II-heat, III-radioactivity, IV-surface tension

119. Four nearly identical gems from the same mineral are weighed using different balances. Their masses are:

3.4533 g, 3.459 g, 3.4656 g, 3.464 g

The four gems are then collected and added to a volumetric cylinder containing 10.00 ml of liquid, and a new volume of 14.97 ml is read. What is the average mass of the four stones and what is the density of the mineral?

A. 3.460 g, and 2.78 g/ml
B. 3.460 g and 2.79 g/ml
C. 3.4605 g and 2.78 g/ml
D. 3.461 g and 2.79 g/ml

120. Which list includes equipment that would not be used in vacuum filtration.

A. Rubber tubing, Florence flask, Büchner funnel
B. Vacuum pump, Hirsch funnel, rubber stopper with a single hole
C. Aspirator, filter paper, filter flask
D. Lab stand, clamp, filter trap

121. Which of the following statements about lab safety is not true?

A. Corrosive chemicals should be stored below eye level.
B. A chemical splash on the eye or skin should be rinsed for 15 minutes in cold water.
C. MSDS means "Material Safety Data Sheet."
D. A student should "stop, drop, and roll" if their clothing catches fire in the lab.

122. Which of the following lists consists entirely of chemicals that are considered safe enough to be in a high school lab?

A. hydrochloric acid, lauric acid, potassium permanganate, calcium hydroxide
B. ethyl ether, nitric acid, sodium benzoate, methanol
C. cobalt (II) sulfide, ethylene glycol, benzoyl peroxide, ammonium chloride
D. picric acid, hydrofluoric acid, cadmium chloride, carbon disulfide.

123. The following procedure was developed to find the specific heat capacity of metals:

1. Place pieces of the metals in an ice-water bath so their initial temperature is 0 °C.
2. Weigh a styrofoam cup.
3. Add water at room temperature to the cup and weigh it again
4. Add a cold metal from the bath to the cup and weigh the cup a third time.
5. Monitor the temperature drop of the water until a final temperature at thermal equilibrium is found.

_____ is also required as additional information in order to obtain heat capacities for the metals. The best control would be to follow the same protocol except to use _____ in step 4 instead of a cold metal.

A. The heat capacity of water / a metal at 100 °C
B. The heat of formation of water / ice from the 0 °C bath
C. The heat of capacity of ice / glass at 0 °C
D. The heat capacity of water / water from the 0 °C bath

124. Which statement about the impact of chemistry on society is not true?

   A. Partial hydrogenation creates *trans* fat.
   B. The Haber Process incorporates nitrogen from the air into molecules for agricultural use.
   C. The CO_2 concentration in the atmosphere has decreased in the last ten years.
   D. The concentration of ozone-destroying chemicals in the stratosphere has decreased in the last ten years.

125. Which statement about everyday applications of chemistry is true?

   A. Rainwater found near sources of air pollution will most likely be basic.
   B. Batteries run down more quickly at low temperatures because chemical reactions are proceeding more slowly.
   C. Benzyl alcohol is a detergent used in shampoo.
   D. Adding salt decreases the time required for water to boil.

Answer Key

1. C	26. D	51. B	76. A	101. C
2. B	27. A	52. D	77. D	102. D
3. D	28. C	53. B	78. C	103. B
4. A	29. C	54. C	79. B	104. B
5. C	30. B	55. D	80. A	105. C
6. D	31. A	56. A	81. D	106. D
7. C	32. A	57. C	82. B	107. C
8. B	33. B	58. C	83. D	108. B
9. A	34. D	59. A	84. B	109. C
10. C	35. B	60. B	85. A	110. A
11. B	36. C	61. D	86. C	111. C
12. D	37. B	62. B	87. A	112. B
13. A	38. D	63. C	88. C	113. B
14. D	39. A	64. B	89. B	114. D
15. B	40. B	65. D	90. D	115. A
16. D	41. C	66. A	91. B	116. C
17. C	42. C	67. B	92. C	117. C
18. B	43. A	68. C	93. A	118. B
19. C	44. D	69. A	94. A	119. B
20. A	45. C	70. D	95. D	120. A
21. C	46. B	71. C	96. C	121. D
22. B	47. D	72. B	97. A	122. A
23. D	48. B	73. A	98. A	123. D
24. A	49. B	74. C	99. B	124. C
25. D	50. A	75. A	100. B	125. B

Rationales with Sample Questions

Note: The first insignificant digit should be carried through intermediate calculations. This digit is shown using *italics* in the solutions below.

1. **A piston compresses a gas at constant temperature. Which gas properties increase?**

   II. **Average speed of molecules**
   III. **Pressure**
   IV. **Molecular collisions with container walls per second**

   A. I and II
   B. I and III
   C. II and III
   D. I, II, and III

C. A decrease in volume (V) occurs at constant temperature (T). Average molecular speed is determined only by temperature and will be constant. V and P are inversely related, so pressure will increase. With less wall area and at higher pressure, more collisions occur per second.

2. **The temperature of a liquid is raised at atmospheric pressure. Which liquid property increases?**

   A. critical pressure
   B. vapor pressure
   C. surface tension
   D. viscosity

B. The critical pressure of a liquid is its vapor pressure at the critical temperature and is always a constant value. A rising temperature increases the kinetic energy of molecules and decreases the importance of intermolecular attraction. More molecules will be free to escape to the vapor phase (vapor pressure increases), but the effect of attractions at the liquid-gas interface will fall (surface tension decreases) and molecules will flow against each other more easily (viscosity decreases).

3. **Potassium crystallizes with two atoms contained in each unit cell. What is the mass of potassium found in a lattice 1.00×10^6 unit cells wide, 2.00×10^6 unit cells high, and 5.00×10^5 unit cells deep?**

    A. 85.0 ng
    B. 32.5 µg
    C. 64.9 µg
    D. 130. µg

D. First we find the number of unit cells in the lattice by multiplying the number in each row, stack, and column:

1.00×10^6 unit cell lengths $\times\ 2.00 \times 10^6$ unit cell lengths $\times\ 5.00 \times 10^5$ unit cell lengths

$= 1.00 \times 10^{18}$ unit cells

Avogadro's number and the molecular weight of potassium (K) are used in the solution:

$$1.00 \times 10^{18}\ \text{unit cells} \times \frac{2\ \text{atoms of K}}{\text{unit cell}} \times \frac{1\ \text{mole of K}}{6.02 \times 10^{23}\ \text{atoms of K}} \times \frac{39.098\ \text{g K}}{1\ \text{mole of K}}$$

$$= 1.30 \times 10^{-4}\ \text{g}$$

$$= 130.\ \text{µg}$$

4. **A gas is heated in a sealed container. Which of the following occur?**

    A. gas pressure rises
    B. gas density decreases
    C. the average distance between molecules increases
    D. all of the above

A. The same material is kept in a constant volume, so neither density nor the distance between molecules will change. Pressure will rise because of increasing molecular kinetic energy impacting container walls.

5. How many molecules are in 2.20 pg of a protein with a molecular weight of 150. kDa?

   A. 8.83×10^9
   B. 1.82×10^9
   C. 8.83×10^6
   D. 1.82×10^6

C. The prefix "p" for "pico-" indicates 10^{-12}. A kilodalton is 1000 atomic mass units.

$$2.20 \text{ pg protein} \times \frac{10^{-12} \text{ g}}{1 \text{ pg}} \times \frac{1 \text{ mole protein}}{150 \times 10^3 \text{ g protein}} \times \frac{6.02 \times 10^{23} \text{ molecules protein}}{1 \text{ mole protein}} =$$
$$= 8.83 \times 10^6 \text{ molecules}$$

6. At STP, 20. μL of O_2 contain 5.4×10^{16} molecules. According to Avogadro's hypothesis, how many molecules are in 20. μL of Ne at STP?

   A. 5.4×10^{15}
   B. 1.0×10^{16}
   C. 2.7×10^{16}
   D. 5.4×10^{16}

D. Avogadro's hypothesis states that equal volumes of different gases at the same temperature and pressure contain equal numbers of molecules.

7. An ideal gas at 50.0 °C and 3.00 atm is in a 300. cm³ cylinder. The cylinder volume changes by moving a piston until the gas is at 50.0 °C and 1.00 atm. What is the final volume?

   A. 100. cm³
   B. 450. cm³
   C. 900. cm³
   D. 1.20 dm³

C. A three-fold decrease in pressure of a constant quantity of gas at constant temperature will cause a three-fold increase in gas volume.

8. Which gas law may be used to solve the previous question?

    A. Charles's law
    B. Boyle's law
    C. Graham's law
    D. Avogadro's law

B. The inverse relationship between volume and pressure is Boyle's law.

9. A blimp is filled with 5000. m^3 of helium at 28.0 °C and 99.7 kPa. What is the mass of helium used?

$$R = 8.3144 \frac{J}{mol\text{-}K}$$

    A. 797 kg
    B. 810. kg
    C. 1.99×10^3 kg
    D. 8.57×10^3 kg

A. First the ideal gas law is manipulated to solve for moles.

$$PV = nRT \quad \Rightarrow \quad n = \frac{PV}{RT}$$

Temperature must be expressed in Kelvin: $T = (28.0 + 273.15)\,K = 301.15\,K$.

The ideal gas law is then used with the knowledge that joules are equivalent to Pa-m^3:

$$n = \frac{PV}{RT} = \frac{\left(99.7 \times 10^3\ Pa\right)\left(5000.\ m^3\right)}{\left(8.3144 \frac{m^3\text{-}Pa}{mol\text{-}K}\right)(301.15\ K)} = 1.991 \times 10^3\ mol\ He.$$

Moles are then converted to grams using the molecular weight of helium:

$$1.991 \times 10^3\ mol\ He \times \frac{4.0026\ g\ He}{1\ mol\ He} = 797 \times 10^3\ g\ He = 797\ kg\ He.$$

10. Which of the following are able to flow from one place to another?

   I. Gases
   II. Liquids
   III. Solids
   IV. Supercritical fluids

   A. I and II
   B. II only
   C. I, II, and IV
   D. I, II, III, and IV

C. Gases and liquids both flow. Supercritical fluids have some traits in common with gases and some in common with liquids, and so they flow also. Solids have a fixed volume and shape.

11. One mole of an ideal gas at STP occupies 22.4 L. At what temperature will one mole of an ideal gas at one atm occupy 31.0 L?

   A. 34.6 °C
   B. 105 °C
   C. 378 °C
   D. 442 °C

B. Either Charles's law, the combined gas law, or the ideal gas law may be used with temperature in Kelvin.

Charles's law or the combined gas law with $P_1 = P_2$ may be manipulated to equate a ratio between temperature and volume when P and n are constant.

$$V \propto T \text{ or } \frac{P_1V_1}{T_1} = \frac{P_2V_2}{T_2} \Rightarrow \frac{T_1}{V_1} = \frac{T_2}{V_2} \Rightarrow T_2 = V_2\frac{T_1}{V_1}$$

$$T_2 = 31.0 \text{ L}\frac{273.15 \text{ K}}{22.4 \text{ L}} = 378 \text{ K} = 105 \text{ °C.}$$

The ideal gas law may also be used with the appropriate gas constant:

$$PV = nRT \Rightarrow T = \frac{PV}{nR}$$

$$T = \frac{(1 \text{ atm})(31.0 \text{ L})}{(1 \text{ mol})\left(0.08206 \frac{\text{L-atm}}{\text{mol-K}}\right)} = 378 \text{ K} = 105 \text{ °C.}$$

12. Why does $CaCl_2$ have a higher normal melting point than NH_3?

  A. London dispersion forces in $CaCl_2$ are stronger than covalent bonds in NH_3.
  B. Covalent bonds in NH_3 are stronger than dipole-dipole bonds in $CaCl_2$.
  C. Ionic bonds in $CaCl_2$ are stronger than London dispersion forces in NH_3.
  D. Ionic bonds in $CaCl_2$ are stronger than hydrogen bonds in NH_3.

D. London dispersion forces are weaker than covalent bonds, eliminating choice A. A higher melting point will result from stronger intermolecular bonds, eliminating choice B. $CaCl_2$ is an ionic solid resulting from a cation on the left and an anion on the right of the periodic table. The dominant attractive forces between NH_3 molecules are hydrogen bonds.

13. Which intermolecular attraction explains the following trend in straight-chain alkanes?

Condensed structural formula	Boiling point (°C)
CH_4	-161.5
CH_3CH_3	-88.6
$CH_3CH_2CH_3$	-42.1
$CH_3CH_2CH_2CH_3$	-0.5
$CH_3CH_2CH_2CH_2CH_3$	36.0
$CH_3CH_2CH_2CH_2CH_2CH_3$	68.7

  A. London dispersion forces
  B. Dipole-dipole interactions
  C. Hydrogen bonding
  D. Ion-induced dipole interactions

A. Alkanes are composed entirely of non-polar C-C and C-H bonds, resulting in no dipole interactions or hydrogen bonding. London dispersion forces increase with the size of the molecule, resulting in a higher temperature requirement to break these bonds and a higher boiling point.

14. List NH_3, PH_3, $MgCl_2$, Ne, and N_2 in order of increasing melting point.

    A. N_2 < Ne < PH_3 < NH_3 < $MgCl_2$
    B. N_2 < NH_3 < Ne < $MgCl_2$ < PH_3
    C. Ne < N_2 < NH_3 < PH_3 < $MgCl_2$
    D. Ne < N_2 < PH_3 < NH_3 < $MgCl_2$

D. Higher melting points result from stronger intermolecular forces. $MgCl_2$ is the only material listed with ionic bonds and will have the highest melting point. Dipole-dipole interactions are present in NH_3 and PH_3 but not in Ne and N_2. Ne and N_2 are also small molecules expected to have very weak London dispersion forces and so will have lower melting points than NH_3 and PH_3. NH_3 will have stronger intermolecular attractions and a higher melting point than PH_3 because hydrogen bonding occurs in NH_3. Ne has a molecular weight of 20 and a spherical shape and N_2 has a molecular weight of 28 and is not spherical. Both of these factors predict stronger London dispersion forces and a higher melting point for N_2. Actual melting points are: Ne (25 K) < N_2 (63 K) < PH_3 (140 K) < NH_3 (195 K) < $MgCl_2$ (987 K).

15. 1-butanol, ethanol, methanol, and 1-propanol are all liquids at room temperature. Rank them in order of increasing viscosity.

    A. 1-butanol < 1-propanol < ethanol < methanol
    B. methanol < ethanol < 1-propanol < 1-butanol
    C. methanol < ethanol < 1-butanol < 1-propanol
    D. 1-propanol < 1-butanol < ethanol < methanol

B. Higher viscosities result from stronger intermolecular attractive forces. The molecules listed are all alcohols with the -OH functional group attached to the end of a straight-chain alkane. In other words, they all have the formula $CH_3(CH_2)_{n-1}OH$. The only difference between the molecules is the length of the alkane corresponding to the value of n. With all else identical, larger molecules have greater intermolecular attractive forces due to a greater molecular surface for the attractions. Therefore the viscosities are ranked: methanol (CH_3OH) < ethanol (CH_3CH_2OH) < 1-propanol ($CH_3CH_2CH_2OH$) < 1-butanol ($CH_3CH_2CH_2CH_2OH$).

16. Which gas has a diffusion rate of 25% the rate for hydrogen?

    A. helium
    B. methane
    C. nitrogen
    D. oxygen

D. Graham's law of diffusion states:

$$\frac{r_1}{r_2} = \sqrt{\frac{M_2}{M_1}}.$$

Hydrogen (H_2) has molecular weight of 2.0158 u. Using the unknown for material #1 and hydrogen for material #2 in the equation for Graham's law, the ratio of rates is:

$$\frac{r_{unknown}}{r_{hydrogen}} = \sqrt{\frac{2.0158\ u}{M_{unknown}}} = 0.25. \quad \text{Squaring both sides yields} \quad \frac{2.0158\ u}{M_{unknown}} = 0.0625.$$

Solving for $M_{unknown}$ gives:

$$M_{unknown} = \frac{2.0158\ u}{0.0625} = 32\ u.$$

The given possibilities are: He (4.0 u), CH_4 (16 u), N_2 (28 u), and O_2 (32 u).

17. 2.00 L of an unknown gas at 1500. mm Hg and a temperature of 25.0 °C weighs 7.52 g. Assuming the ideal gas equation, what is the molecular weight of the gas?

$$760 \text{ mm Hg} = 1 \text{ atm}$$
$$R = 0.08206 \text{ L-atm/(mol-K)}$$

A. 21.6 u
B. 23.3 u
C. 46.6 u
D. 93.2 u

C. Pressure and temperature must be expressed in the proper units. Next the ideal gas law is used to find the number of moles of gas.

$$P = 1500 \text{ mm Hg} \times \frac{1 \text{ atm}}{760 \text{ mm Hg}} = 1.974 \text{ atm and } T = 25.0 + 273.15 = 298.15 \text{ K}$$

$$PV = nRT \implies n = \frac{PV}{RT}$$

$$n = \frac{(1.974 \text{ atm})(2.00 \text{ L})}{\left(0.08206 \frac{\text{L-atm}}{\text{mol-K}}\right)(298.15 \text{ K})} = 0.1613 \text{ mol.}$$

The molecular mass may be found from the mass of one mole.

$$\frac{7.52 \text{ g}}{0.1613 \text{ mol}} = 46.6 \frac{\text{g}}{\text{mol}} \implies 46.6 \text{ u}$$

18. Which substance is most likely to be a gas at STP?

A. SeO_2
B. F_2
C. $CaCl_2$
D. I_2

B. A gas at STP has a normal boiling point under 0 °C. The substance with the lowest boiling point will have the weakest intermolecular attractive forces and will be the most likely gas at STP. F_2 has the lowest molecular weight, is not a salt, metal, or covalent network solid, and is non-polar, indicating the weakest intermolecular attractive forces of the four choices. F_2 actually is a gas at STP, and the other three are solids.

19. What pressure is exerted by a mixture of 2.7 g of H2 and 59 g of Xe at 0 °C on a 50. L container?

A. 0.69 atm
B. 0.76 atm
C. 0.80 atm
D. 0.97 atm

C. Grams of gas are first converted to moles:

$$2.7 \text{ g H}_2 \times \frac{1 \text{ mol H}_2}{2 \times 1.0079 \text{ g H}_2} = 1.33 \text{ mol H}_2 \quad \text{and} \quad 59 \text{ g Xe} \times \frac{1 \text{ mol H}_2}{131.29 \text{ g H}_2} = 0.449 \text{ mol Xe}$$

Dalton's law of partial pressures for an ideal gas is used to find the pressure of the mixture:

$$P_{total}V = \left(n_{H_2} + n_{Xe}\right)RT \Rightarrow P_{total} = \frac{\left(n_{H_2} + n_{Xe}\right)RT}{V}$$

$$P_{total} = \frac{(1.33 \text{ mol} + 0.449 \text{ mol})\left(0.08206 \frac{\text{L-atm}}{\text{mol-K}}\right)(273.15 \text{ K})}{50. \text{ L}} = 0.80 \text{ atm.}$$

20. A few minutes after opening a bottle of perfume, the scent is detected on the other side of the room. What law relates to this phenomenon?

A. Graham's law
B. Dalton's law
C. Boyle's law
D. Avogadro's law

A. Graham's law describes the rate of diffusion (or effusion) of a gas, in this instance, the rate of diffusion of molecules in perfume vapor.

21. Which of the following statements are true of vapor pressure at equilibrium?

    A. Solids have no vapor pressure.
    B. Dissolving a solute in a liquid increases its vapor pressure.
    C. The vapor pressure of a pure substance is characteristic of that substance and its temperature.
    D. All of the above

C. Only temperature and the identity of the substance determine vapor pressure. Solids have a vapor pressure, and solutes decrease vapor pressure.

22. Find the partial pressure of N_2 in a container holding H_2O and N_2 at 150. kPa and 50 °C. The vapor pressure of H_2O at 50 °C is 12 kPa.

    A. 12 kPa
    B. 138 kPa
    C. 162 kPa
    D. The value cannot be determined.

B. The partial pressure of H_2O vapor in the container is its vapor pressure. The partial pressure of N_2 may be found by manipulating Dalton's law:
$$P_{total} = P_{H_2O} + P_{N_2} \Rightarrow P_{N_2} = P_{total} - P_{H_2O}$$
$$P_{N_2} = P_{total} - P_{H_2O} = 150. \text{ kPa} - 12 \text{ kPa} = 138 \text{ kPa}$$

23. The normal boiling point of water on the Kelvin scale is closest to:

    A. 112 K
    B. 212 K
    C. 273 K
    D. 373 K

D. Temperature in Kelvin are equal to Celsius temperatures plus 273.15. Since the normal boiling point of water is 100 °C, it will boil at 373.15 K, corresponding to answer D.

24. Which phase may be present at the triple point of a substance?

I.   Gas
II. Liquid
III. Solid
IV. Supercritical fluid

A. I, II, and III
B. I, II, and IV
C. II, III, and IV
D. I, II, III, and IV

A. Gas, liquid and solid may exist together at the triple point.

25. In the following phase diagram, _____ occurs as P is decreased from A to B at constant T and _____ occurs as T is increased from C to D at constant P.

A. deposition, melting
B. sublimation, melting
C. deposition, vaporization
D. sublimation, vaporization

D. Point A is located in the solid phase, point C is located in the liquid phase. Points B and D are located in the gas phase. The transition from solid to gas is sublimation and the transition from liquid to gas is vaporization.

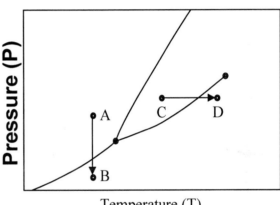

26. Heat is added to a pure solid at its melting point until it all becomes liquid at its freezing point. Which of the following occur?

A. Intermolecular attractions are weakened.
B. The kinetic energy of the molecules does not change.
C. The freedom of the molecules to move about increases.
D. All of the above

D. Intermolecular attractions are lessened during melting. This permits molecules to move about more freely, but there is no change in the kinetic energy of the molecules because the temperature has remained the same.

27. Which of the following occur when NaCl dissolves in water?

    A. Heat is required to break bonds in the NaCl crystal lattice.
    B. Heat is released when hydrogen bonds in water are broken.
    C. Heat is required to form bonds of hydration.
    D. The oxygen end of the water molecule is attracted to the Cl^- ion.

A. The lattice does break apart, H-bonds in water are broken, and bonds of hydration are formed, but the first and second process require heat while the third process releases heat. The oxygen end of the water molecule has a partial negative charge and is attracted to the Na^+ ion.

28. The solubility of $CoCl_2$ is 54 g per 100 g of ethanol. Three flasks each contain 100 g of ethanol. Flask #1 also contains 40 g $CoCl_2$ in solution. Flask #2 contains 56 g $CoCl_2$ in solution. Flask #3 contains 5 g of solid $CoCl_2$ in equilibrium with 54 g $CoCl_2$ in solution. Which of the following describe the solutions present in the liquid phase of the flasks?

    A. #1-saturated, #2-supersaturated, #3-unsaturated.
    B. #1-unsaturated, #2-miscible, #3-saturated.
    C. #1-unsaturated, #2-supersaturated, #3-saturated.
    D. #1-unsaturated, #2-not at equilibrium, #3-miscible.

C. Flask #1 contains less solute than the solubility limit, and is unsaturated. Flask #2 contains more solute than the solubility limit, and is supersaturated and also not at equilibrium. Flask #3 contains the solubility limit and is a saturated solution. The term "miscible" applies only to liquids that mix together in all proportions.

29. The solubility at 1.0 atm of pure CO_2 in water at 25 °C is 0.034 M. According to Henry's law, what is the solubility at 4.0 atm of pure CO_2 in water at 25 °C? Assume no chemical reaction occurs between CO_2 and H_2O.

    A. 0.0085 M
    B. 0.034 M
    C. 0.14 M
    D. 0.25 M

C. Henry's law states that CO_2 solubility in M (mol/L) will be proportional to the partial pressure of the gas. A four-fold increase in pressure from 1.0 atm to 4.0 atm will increase solubility four-fold from 0.034 M to 0.14 M.

30. Carbonated water is bottled at 25 °C under pure CO_2 at 4.0 atm. Later the bottle is opened at 4 °C under air at 1.0 atm that has a partial pressure of 3×10^{-4} atm CO_2. Why do CO_2 bubbles form when the bottle is opened?

   A. CO_2 leaves the solution due to a drop in solubility at the lower total pressure.
   B. CO_2 leaves the solution due to a drop in solubility at the lower CO_2 pressure.
   C. CO_2 leaves the solution due to a drop in solubility at the lower temperature.
   D. CO_2 is formed by the decomposition of carbonic acid.

B. A is incorrect because if the water were bottled under a different gas at a high pressure, it would not be carbonated. CO_2 partial pressure is the important factor in solubility. C is incorrect because a decrease in temperature will increase solubility, and the chance from 298 K to 277 K is relatively small. D may occur, but this represents a small fraction of the gas released.

31. When KNO_3 dissolves in water, the water grows slightly colder. An increase in temperature will _____ the solubility of KNO_3.

   A. increase
   B. decrease
   C. have no effect on
   D. have an unknown effect with the information given on

A. The decline in water temperature indicates that the net solution process is endothermic (requiring heat). A temperature increase supplying more heat will favor the solution and increase solubility according to Le Chatelier's principle.

32. An experiment requires 100. mL of a 0.500 M solution of $MgBr_2$. How many grams of $MgBr_2$ will be present in this solution?

   A. 9.21 g
   B. 11.7 g
   C. 12.4 g
   D. 15.6 g

A.

$$0.100 \text{ L solution} \times \frac{0.500 \text{ mol MgBr}_2}{L} \times \frac{(24.305 + 2 \times 79.904) \text{ g MgBr}_2}{\text{mol MgBr}_2} = 9.21 \text{ g MgBr}_2$$

33. 500. mg of RbOH are added to 500. g of ethanol (C_2H_6O) resulting in 395 mL of solution. Determine the molarity and molality of RbOH.

   A. 0.0124 M, 0.00488 m
   B. 0.0124 M, 0.00976 m
   C. 0.0223 M, 0.00488 m
   D. 0.0223 M, 0. 00976 m

B. First we determine the moles of solute present:

$$0.500 \text{ g RbOH} \times \frac{1 \text{ mol RbOH}}{(85.468 + 15.999 + 1.0079) \text{ g RbOH}} = 0.004879 \text{ mol RbOH}.$$

This value is used to calculate molarity and molality:

$$\frac{0.04879 \text{ mol RbOH}}{0.395 \text{ L solution}} = 0.0124 \text{ M RbOH}$$

and

$$\frac{0.04879 \text{ mol RbOH}}{0.500 \text{ kg ethanol}} = 0.00976 \ m \text{ RbOH}.$$

34. **20.0 g H₃PO₄ in 1.5 L of solution are intended to react with KOH according to the following reaction:** $H_3PO_4 + 3\,KOH \rightarrow K_3PO_4 + 3\,H_2O$. **What is the molarity and normality of the H₃PO₄ solution?**

A. 0.41 M, 1.22 N
B. 0.41 M, 0.20 N
C. 0.14 M, 0.045 N
D. 0.14 M, 0. 41 N

D. We use two methods to solve this problem. In the first method, we determine the moles of solute present and use it to calculate molarity and normality:

$$20.0 \text{ g } H_3PO_4 \times \frac{1 \text{ mol } H_3PO_4}{(3 \times 1.0079 + 30.974 + 4 \times 15.999) \text{ g } H_3PO_4} = 0.204 \text{ mol } H_3PO_4.$$

$$\frac{0.204 \text{ mol } H_3PO_4}{1.5 \text{ L solution}} = 0.136 \frac{\text{mol } H_3PO_4}{L} = 0.14 \text{ M } H_3PO_4$$

and

$$\frac{0.204 \text{ mol } H_3PO_4}{1.5 \text{ L solution}} \times \frac{3 \text{ reaction equivalents}}{1 \text{ mol } H_3PO_4} = 0.408 \frac{\text{reaction equivalents}}{L} = 0.41 \text{ N } H_3PO_4.$$

Alternatively, molarity may be found in one step and normality may be determined from the molarity:

$$\frac{20.0 \text{ g } H_3PO_4}{1.5 \text{ L}} \times \frac{1 \text{ mol } H_3PO_4}{(3 \times 1.0079 + 30.974 + 4 \times 15.999) \text{ g } H_3PO_4} = 0.136 \frac{\text{mol } H_3PO_4}{L} = 0.14 \text{ M } H_3PO_4$$

and

$$0.136 \frac{\text{mol } H_3PO_4}{L} \times \frac{3 \text{ reaction equivalents}}{1 \text{ mol } H_3PO_4} = 0.408 \frac{\text{reaction equivalents}}{L} = 0.41 \text{ N.}$$

35. Aluminum sulfate is a strong electrolyte. What is the concentration of all species in a 0.2 M solution of aluminum sulfate?

A. 0.2 M Al^{3+}, 0.2 M SO_4^{2-}
B. 0.4 M Al^{3+}, 0.6 M SO_4^{2-}
C. 0.6 M Al^{3+}, 0.4 M SO_4^{2-}
D. 0.2 M $Al_2(SO_4)_3$

B. A strong electrolyte will completely ionize into its cation and anion. Aluminum sulfate is $Al_2(SO_4)_3$. Each mole of aluminum sulfate ionizes into 2 moles of Al^{3+} and 3 moles of SO_4^{2-}:

$$0.2 \frac{\text{mol } Al_2(SO_4)_3}{L} \times \frac{2 \text{ mol } Al^{3+}}{\text{mol } Al_2(SO_4)_3} = 0.4 \frac{\text{mol } Al^{3+}}{L} \text{ and}$$

$$0.2 \frac{\text{mol } Al_2(SO_4)_3}{L} \times \frac{3 \text{ mol } SO_4^{2-}}{\text{mol } Al_2(SO_4)_3} = 0.6 \frac{\text{mol } SO_4^{2-}}{L}.$$

36. 15 g of formaldehyde (CH_2O) are dissolved in 100. g of water. Calculate the weight percentage and mole fraction of formaldehyde in the solution.

A. 13%, 0.090
B. 15%, 0.090
C. 13%, 0.083
D. 15%, 0.083

C. Remember to use the total amounts in the denominator.

For weight percentage: $\dfrac{15 \text{ g } CH_2O}{(15+100) \text{ g total}} = 0.13 = 13\%$.

For mole fraction, first convert grams of each substance to moles:

$$15 \text{ g } CH_2O \times \frac{\text{mol } CH_2O}{(12.011+2 \times 1.0079+15.999) \text{ g } CH_2O} = 0.4996 \text{ mol } CH_2O$$

$$100 \text{ g } H_2O \times \frac{\text{mol } H_2O}{(2 \times 1.0079+15.999) \text{ g } H_2O} = 5.551 \text{ mol } H_2O.$$

Again use the total amount in the denominator $\dfrac{0.4996 \text{ mol } CH_2O}{(0.4996 + 5.551) \text{ mol total}} = 0.083$.

37. Which of the following would make the best solvent for Br_2?

    A. H_2O
    B. CS_2
    C. NH_3
    D. molten NaCl

B. The best solvents for a solute have intermolecular bonds of similar strength to the solute ("like dissolves like"). Bromine is a non-polar molecule with intermolecular attractions due to weak London dispersion forces. The relatively strong hydrogen bonding in H_2O and NH_3 and the very strong electrostatic attractions in molten NaCl would make each of them a poor solvent for Br_2 because these molecules would prefer to remain attracted to one another. CS_2 is a fairly small non-polar molecule.

38. Which of the following is most likely to dissolve in water?

    A. H_2
    B. CCl_4
    C. $(SiO_2)_n$
    D. CH_3OH

D. The best solutes for a solvent have intermolecular bonds of similar strength to the solvent. H_2O molecules are connected by fairly strong hydrogen bonds. H_2 and CCl_4 are molecules with intermolecular attractions due to weak London dispersion forces. $(SiO_2)_n$ is a covalent network solid and is essentially one large molecule with bonds that much stronger than hydrogen bonds. CH_3OH (methanol) is miscible with water because it contains hydrogen bonds between molecules.

39. Which of the following is <u>not</u> a colligative property?

    A. Viscosity lowering
    B. Freezing point lowering
    C. Boiling point elevation
    D. Vapor pressure lowering

A. Vapor pressure lowering, boiling point elevation, and freezing point lowering may all be visualized as a result of solute particles interfering with the interface between phases in a consistent way. This is not the case for viscosity.

40. $BaCl_2(aq) + Na_2SO_4(aq) \rightarrow BaSO_4(s) + 2NaCl(aq)$ is an example of a _____ reaction.

    A. acid-base
    B. precipitation
    C. redox
    D. nuclear

B. $BaSO_4$ falls out of the solution as a precipitate, but the charges on Ba^{2+} and SO_4^{2-} remain unchanged, so this is not a redox reaction. Neither $BaCl_2$ nor Na_2SO_4 are acids or bases, and the nuclei involved also remain unaltered

41. **List the following aqueous solutions in order of increasing boiling point.**

    I.   0.050 *m* $AlCl_3$
    II. 0.080 *m* $Ba(NO_3)_2$
    III. 0.090 *m* NaCl
    IV. 0.12 *m* ethylene glycol ($C_2H_6O_2$)

    A. I < II < III < IV
    B. I < III < IV < II
    C. IV < III < I < II
    D. IV < III < II < I

C. Particles in solution determine colligative properties. The first three materials are strong electrolyte salts, and $C_2H_6O_2$ is a non-electrolyte.

$AlCl_3(aq)$ is $Al^{3+} + 3\,Cl^-$. So $0.050 \dfrac{\text{mol } AlCl_3}{\text{kg } H_2O} \times \dfrac{4 \text{ mol particles}}{\text{mol } AlCl_3} = 0.200\ m$ particles

$Ba(NO_3)_2(aq)$ is $Ba^{2+} + 2\,NO_3^-$. So $0.080 \dfrac{\text{mol } Ba(NO_3)_2}{\text{kg } H_2O} \times \dfrac{3 \text{ mol particles}}{\text{mol } Ba(NO_3)_2} = 0.240\ m$ particles

$NaCl(aq)$ is $Na^+ + Cl^-$. So $0.090 \dfrac{\text{mol } NaCl}{\text{kg } H_2O} \times \dfrac{2 \text{ mol particles}}{\text{mol } NaCl} = 0.180\ m$ particles

$C_2H_6O_2(aq)$ is not an electrolyte. So $0.12 \dfrac{\text{mol } C_2H_6O_2}{\text{kg } H_2O} \times \dfrac{1 \text{ mol particles}}{\text{mol } C_2H_6O_2} = 0.12\ m$ particles

The greater the number of dissolved particles, the greater the boiling point elevation.

42. Osmotic pressure is the pressure required to prevent _____ flowing from low to high _____ concentration across a semipermeable membrane.

   A. solute, solute
   B. solute, solvent
   C. solvent, solute
   D. solvent, solvent

C. Osmotic pressure is the pressure required to prevent osmosis, which is the flow of solvent across the membrane from low to high solute concentration. This is also the direction from high to low solvent concentration.

43. A solution of NaCl in water is heated on a mountain in an open container until it boils at 100. °C. The air pressure on the mountain is 0.92 atm. According to Raoult's law, what mole fraction of Na^+ and Cl^- are present in the solution?

   A. 0.04 Na^+, 0.04 Cl^-
   B. 0.08 Na^+, 0.08 Cl^-
   C. 0.46 Na^+, 0.46 Cl^-
   D. 0.92 Na^+, 0.92 Cl^-

A. The vapor pressure of H_2O at 100. °C is exactly 1 atm. Boiling point decreases with external pressure, so the boiling point of pure H_2O at 0.9 atm will be less than 100. °C. Adding salt raises the boiling point at 0.92 atm to 100. °C by decreasing vapor pressure to 0.92 atm. According to Raoult's law:

$$P^{vapor}_{solution} = P^{vapor}_{pure\ solvent}\left(mole\ fraction\right)_{solvent} \Rightarrow \left(mole\ fraction\right)_{solvent} = \frac{P^{vapor}_{solution}}{P^{vapor}_{pure\ solvent}}.$$

$$\text{Therefore, } \left(mole\ fraction\right)_{H_2O} = \frac{0.92\ atm\ at\ 100.\ °C}{1.0\ atm\ at\ 100.\ °C} = 0.92\ \frac{mol\ H_2O}{mol\ total}.$$

The remaining 0.08 mole fraction of solute is evenly divided between the two ions:

$$\left(mole\ fraction\right)_{solute} = 1 - \left(mole\ fraction\right)_{H_2O} = 1 - 0.92 = 0.08\ \frac{mol\ solute\ particles}{mol\ total}$$

$$\left(mole\ fraction\right)_{Na^+} = 0.08\ \frac{mol\ solute\ particles}{mol\ total} \times \frac{1\ mol\ Na^+}{2\ mol\ solute\ particles} = 0.04\ \frac{mol\ Na^+}{mol\ total}$$

$$\left(mole\ fraction\right)_{Cl^-} = 0.08\ \frac{mol\ solute\ particles}{mol\ total} \times \frac{1\ mol\ Cl^-}{2\ mol\ solute\ particles} = 0.04\ \frac{mol\ Cl^-}{mol\ total}.$$

44. Write a balanced nuclear equation for the emission of an alpha particle by polonium-209.

    A. $^{209}_{84}Po \rightarrow {}^{205}_{81}Pb + {}^{4}_{2}He$

    B. $^{209}_{84}Po \rightarrow {}^{205}_{82}Bi + {}^{4}_{2}He$

    C. $^{209}_{84}Po \rightarrow {}^{209}_{85}At + {}^{0}_{-1}e$

    D. $^{209}_{84}Po \rightarrow {}^{205}_{82}Pb + {}^{4}_{2}He$

D. The periodic table before skill 1.1 shows that polonium has an atomic number of 84. The emission of an alpha particle, $^{4}_{2}He$ (eliminating choice C), will leave an atom with an atomic number of 82 and a mass number of 205 (eliminating choice A). The periodic table identifies this element as lead, $^{205}_{82}Pb$, not bismuth (eliminating choice B).

45. Write a balanced nuclear equation for the decay of calcium-45 to scandium-45.

    A. $^{45}_{20}Ca \rightarrow {}^{41}_{18}Sc + {}^{4}_{2}He$

    B. $^{45}_{20}Ca + {}^{0}_{1}e \rightarrow {}^{45}_{21}Sc$

    C. $^{45}_{20}Ca \rightarrow {}^{45}_{21}Sc + {}^{0}_{-1}e$

    D. $^{45}_{20}Ca + {}^{0}_{1}p \rightarrow {}^{45}_{21}Sc$

C. All four choices are balanced mathematically. "A" leaves scandium-41 as a decay product, not scandium-45. "B" and "D" require the addition of particles not normally present in the atom. If these reactions do occur, they are not decay reactions because they are not spontaneous. "C" involves the common decay mechanism of beta emission.

46. $^{3}_{1}H$ decays with a half-life of 12 years. 3.0 g of pure $^{3}_{1}H$ were placed in a sealed container 24 years ago. How many grams of $^{3}_{1}H$ remain?

    A. 0.38 g
    B. 0.75 g
    C. 1.5 g
    D. 3.0 g

B. Every 12 years, the amount remaining is cut in half. After 12 years, 1.5 g will remain. After another 12 years, 0.75 g will remain.

47. Oxygen-15 has a half-life of 122 seconds. What percentage of a sample of oxygen-15 has decayed after 300. seconds?

    A. 18.2%
    B. 21.3%
    C. 78.7%
    D. 81.8%

D. We may assume a convenient number (like 100.0 g) for a sample size. The amount remaining may be found from:

$$A_{remaining} = A_{initially} \left(\frac{1}{2}\right)^{\frac{t}{t_{halflife}}}$$

$$= 100.0 \text{ g } {}^{15}O \left(\frac{1}{2}\right)^{\frac{300. \text{ seconds}}{122 \text{ seconds}}} = 18.2 \text{ g } {}^{15}O$$

We are asked to determine the percentage that has decayed. This will be 100.0 g – 18.2 g = 81.8 g or 81.8% of the initial sample.

48. Which of the following isotopes is commonly used for medical imaging in the diagnose of diseases?

    A. cobalt-60
    B. technetium-99m
    C. tin-117m
    D. plutonium-238

B. The other three isotopes have limited medical applications (tin-117m has been used for the relief of bone cancer pain), but only Tc-99m is used routinely for imaging.

49. Carbon-14 dating would be useful in obtaining the age of which object?

    A. a 20th century Picasso painting
    B. a mummy from ancient Egypt
    C. a dinosaur fossil
    D. all of the above

B. C-14 is used in archeology because its half-life is 5730 years. Too little C-14 would have decayed from the painting and nearly all of the C-14 would have decayed from the fossil. In both cases, an estimate of age would be impossible with this isotope.

50. Which of the following isotopes can create a chain reaction of nuclear fission?

   A. uranium-235
   B. uranium-238
   C. plutonium-238
   D. all of the above

A. Uranium-235 and plutonium-239 are the two fissile isotopes used for nuclear power. ^{238}U is the most common uranium isotope. ^{238}Pu is used as a heat source for energy in space probes and some pacemakers.

51. List the following scientists in chronological order from earliest to most recent with respect to their most significant contribution to atomic theory:

   I. John Dalton
   II. Niels Bohr
   III. J. J. Thomson
   IV. Ernest Rutherford

   A. I, III, II, IV
   B. I, III, IV, II
   C. I, IV, III, II
   D. III, I, II, IV

B. Dalton founded modern atomic theory. J.J. Thomson determined that the electron is a subatomic particle but he placed it in the center of the atom. Rutherford discovered that electrons surround a small dense nucleus. Bohr determined that electrons may only occupy discrete positions around the nucleus.

52. Match the theory with the scientist who first proposed it:

    I.   Electrons, atoms, and all objects with momentum also exist as waves.

    II. Electron density may be accurately described by a single mathematical equation.

    III. There is an inherent indeterminacy in the position and momentum of particles.

    IV. Radiant energy is transferred between particles in exact multiples of a discrete unit.

    A. I-de Broglie, II-Planck, III-Schrödinger, IV-Thomson
    B. I-Dalton, II-Bohr, III-Planck, IV-de Broglie
    C. I-Henry, II-Bohr, III-Heisenberg, IV-Schrödinger
    D. I-de Broglie, II-Schrödinger, III-Heisenberg, IV-Planck

    D. Henry's law relates gas partial pressure to liquid solubility.

53. How many neutrons are in $^{60}_{27}$Co ?

    A. 27
    B. 33
    C. 60
    D. 87

    B. The number of neutrons is found by subtracting the atomic number (27) from the mass number (60).

54. The terrestrial composition of an element is: 50.7% as an isotope with an atomic mass of 78.9 u and 49.3% as an isotope with an atomic mass of 80.9 u. Both isotopes are stable. Calculate the atomic mass of the element.

    A. 79.0 u
    B. 79.8 u
    C. 79.9 u
    D. 80.8 u

    C.

Atomic mass of element = (Fraction as 1st isotope) (Atomic mass of 1st isotope)

+

(Fraction as 2nd isotope) (Atomic mass of 2nd isotope)

$= (0.507) (78.9 \text{ u}) + (0.493) (80.9 \text{ u}) = 79.89 \text{ u} = 79.9 \text{ u}$

55. Which of the following is a correct electron arrangement for oxygen?

A.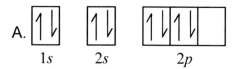

B. $1s^2 1p^2 2s^2 2p^2$
C. 2, 2, 4
D. none of the above

D. Choice A violates Hund's rule. The two electrons on the far right should occupy the final two orbitals. B should be $1s^2 2s^2 2p^4$. There is no $1p$ subshell. C should be 2, 6. Number lists indicate electrons in shells.

56. Which of the following statements about radiant energy is <u>not</u> true?

A. The energy change of an electron transition is directly proportional to the wavelength of the emitted or absorbed photon.
B. The energy of an electron in a hydrogen atom depends only on the principle quantum number.
C. The frequency of photons striking a metal determines whether the photoelectric effect will occur.
D. The frequency of a wave of electromagnetic radiation is inversely proportional to its wavelength.

A. The energy change (ΔE) is <u>inversely</u> proportional to the wavelength (λ) of the photon according the equations:

$$\Delta E = \frac{hc}{\lambda}.$$

where h is Planck's constant and c is the speed of light.

Choice B is true for hydrogen. Atoms with more than one electron are more complex. The frequency of individual photons, not the number of photons determines whether the photoelectric effect occurs, so choice C is true. Choice D is true. The proportionality constant is the speed of light according to the equation:

$$v = \frac{c}{\lambda}.$$

57. Match the orbital diagram for the ground state of carbon with the rule/principle it violates:

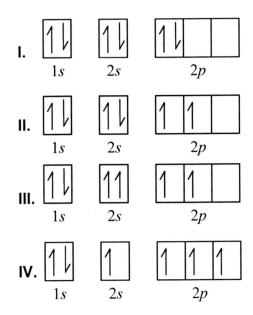

A. I-Pauli exclusion, II-Aufbau, III-no violation, IV-Hund's
B. I-Aufbau, II-Pauli exclusion, III-no violation, IV-Hund's
C. I-Hund's, II-no violation, III-Pauli exclusion, IV-Aufbau
D. I-Hund's, II-no violation, III-Aufbau, IV-Pauli exclusion

C. Diagram I violates Hund's rule because a second electron is added to a degenerate orbital before all orbitals in the subshell have one electron. Diagram III violates the Pauli exclusion principle because both electrons in the 2s orbital have the same spin. They would have the same 4 quantum numbers. Diagram IV violates the Aufbau principle because an electron occupies the higher energy 2p orbital before 2s orbital has been filled; this configuration is not at the ground state.

58. Select the list of atoms that are arranged in order of increasing size.

A. Mg, Na, Si, Cl
B. Si, Cl, Mg, Na
C. Cl, Si, Mg, Na
D. Na, Mg, Si, Cl

C. These atoms are all in the same row of the periodic table. Size increases further to the left for atoms in the same row.

59. Based on trends in the periodic table, which of the following properties would you expect to be greater for Rb than for K?

    I. Density
    II. Melting point
    III. Ionization energy
    IV. Oxidation number in a compound with chlorine

    A. I only
    B. I, II, and III
    C. II and III
    D. I, II, III, and IV

A. Rb is underneath K in the alkali metal column (group 1) of the periodic table. There is a general trend for density to increase lower on the table for elements in the same row, so we select choice I. Rb and K experience metallic bonds for intermolecular forces and the strength of metallic bonds decreases for larger atoms further down the periodic table resulting in a lower melting point for Rb, so we do not choose II. Ionization energy decreases for larger atoms further down the periodic table, so we do not choose III. Both Rb and K would be expected to have a charge of +1 and therefore an oxidation number of +1 in a compound with chlorine, so we do not choose IV.

60. Which oxide forms the strongest acid in water?

    A. Al_2O_3
    B. Cl_2O_7
    C. As_2O_5
    D. CO_2

B. The strength of acids formed from oxides increases with electronegativity and with oxidation state. We know Cl has a greater electronegativity than Al, As, and C because it is closer to the top right of the periodic table. The oxidation numbers of our choices are +3 for Al, +7 for Cl, +5 for As, and +4 for C. Both its electronegativity and its oxidation state indicate Cl_2O_7 will form the strongest acid.

61. Rank the following bonds from least to most polar:

C-H, C-Cl, H-H, C-F

 A. C-H < H-H < C-F < C-Cl
 B. H-H < C-H < C-F < C-Cl
 C. C-F < C-Cl < C-H < H-H
 D. H-H < C-H < C-Cl < C-F

D. Bonds between atoms of the same element are completely non-polar, so H-H is the least polar bond in the list, eliminating choices A and C. The C-H bond is considered to be non-polar even though the electrons of the bond are slightly unequally shared. C-Cl and C-F are both polar covalent bonds, but C-F is more strongly polar because F has a greater electronegativity.

62. At room temperature, CaBr$_2$ is expected to be:

 A. a ductile solid
 B. a brittle solid
 C. a soft solid
 D. a gas

B. Ca is a metal because it is on the left of the periodic table, and Br is a non-metal because it is on the right. The compound they form together will be an ionic salt, and ionic salts are brittle solids (choice B) at room temperature. NaCl is another example.

63. Which of the following is a proper Lewis dot structure of CHClO?

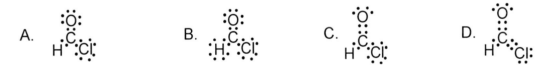

C. C has 4 valence shell electrons, H has 1, Cl has 7, and O has 6. The molecule has a total of 18 valence shell electrons. This eliminates choice B which has 24. Choice B is also incorrect because has an octet around a hydrogen atom instead of 2 electrons and because there are only six electrons surrounding the central carbon. A single bond connecting all atoms would give choice A. This is incorrect because there are only 6 electrons surrounding the central carbon. A double bond between C and O gives the correct answer, C. A double bond between C and O and also between C and Cl would give choice D. This is incorrect because there are 10 electrons surrounding the central carbon.

64. In C_2H_2, each carbon atom contains the following valence orbitals:

A. *p* only
B. *p* and *sp* hybrids
C. *p* and sp^2 hybrids
D. sp^3 hybrids only

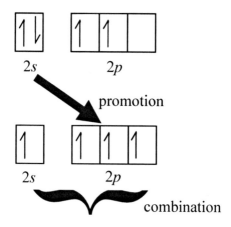

2*s*   2*p*

promotion

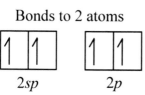

2*s*   2*p*

combination

B. An isolated C has the valence electron configuration $2s^22p^2$. Before bonding, one *s* electron is promoted to an empty *p* orbital. In C_2H_2, each C atom bonds to 2 other atoms. Bonding to two other atoms is achieved by combination into two *p* orbitals and two *sp* hybrids.

Bonds to 2 atoms

2*sp*   2*p*

65. Which statement about molecular structures is false?

A. 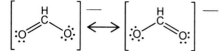   is a conjugated molecule.

B. A bonding σ orbital connects two atoms by the straight line between them.
C. A bonding π orbital connects two atoms in a separate region from the straight line between them.
D. The anion with resonance forms will always exist in one form or the other.

D. A conjugated molecule is a molecule with double bonds on adjacent atoms such as the molecule shown in A. Choice B and C give the definition of sigma and pi molecular orbitals. D is false because a resonance form is one of multiple equivalent Lewis structures, but these structures do not describe the actual state of the molecule. The anion will exist in a state between the two forms.

66. What is the shape of the PH_3 molecule? Use the VSEPR model.

    A. Trigonal pyramidal
    B. Trigonal bipyramidal
    C. Trigonal planar
    D. Tetrahedral

A. The Lewis structure for PH_3 is given to the right. This structure contains 4 electron pairs around the central atom, so the geometral arrangement is tetrahedral. However, the shape of a molecule is given by its atom locations, and there are only three atoms so choice D is not correct. Four electrons pairs with one unshared pair (3 bonds and one lone pair) give a trigonal pyramidal shape as shown to the left.

67. What is the chemical composition of magnesium nitrate?

    A. 11.1% Mg, 22.2% N, 66.7% O
    B. 16.4% Mg, 18.9% N, 64.7% O
    C. 20.9% Mg, 24.1% N, 55.0% O
    D. 28.2% Mg, 16.2% N, 55.7% O

B. First find the formula for magnesium nitrate. Mg is an alkali earth metal (**Skill 3.1**) and will always have a 2+ charge. The nitrate ion is NO_3^- (**Skill 5.2**). Two nitrate ions are required for each Mg^{2+} ion. Therefore the formula is $Mg(NO_3)_2$

    Skill 5.1 describes determination chemical composition.
      1) Determine the number of atoms for elements in $Mg(NO_3)_2$: 1 Mg, 2 N, 6 O.
      2) Multiply by the molecular weight of the elements to determine the grams of each in one mole of the formula.

$$\frac{1 \text{ mol Mg}}{\text{mol Mg(NO}_3)_2} \times \frac{24.3 \text{ g Mg}}{\text{mol Mg}} = 24.3 \text{ g Mg/mol Mg(NO}_3)_2$$

$$2(14.0) = 28.0 \text{ g N/mol Mg(NO}_3)_2$$

$$6(16.0) = 96.0 \text{ g O/mol Mg(NO}_3)_2$$

$$\overline{148.3 \text{ g Mg(NO}_3)_2/\text{mol Mg(NO}_3)_2}$$

      3) Determine formula mass
      4) Divide to determine % composition

$$\%Mg = \frac{24.3 \text{ g Mg/mol Mg(NO}_3)_2}{148.3 \text{ g Mg(NO}_3)_2/\text{mol Mg(NO}_3)_2} = 0.164 \text{ g Mg/g Mg(NO}_3)_2 \times 100\% = 16.4\%$$

$$\%N = \frac{28.0}{148.3} \times 100\% = 18.9\% \qquad \%O = \frac{96.0}{148.3} \times 100\% = 64.7\%$$

Answer A is the fractional representation of the presence of each atom in the formula. Composition is based on mass percentage. Answer C is the chemical composition of $Mg(NO_2)_2$, magnesium nitrite. Answer D is the chemical composition of "$MgNO_3$", a formula that results from not balancing charges.

68. The IUPAC name for Cu_2SO_3 is:

A. Dicopper sulfur trioxide
B. Copper (II) sulfate
C. Copper (I) sulfite
D. Copper (II) sulfite

C. Cu_2SO_3 is an ionic compound containing copper cation and the SO_3 anion. Choice A is wrong because it uses the naming system for molecular compounds. The SO_3 anion is 2– and is named sulfite. It takes two copper cations in to neutralize this charge, so Cu has a charge of 1+, and the name is copper (I) sulfite.

69. Which name or formula is not represented properly?

A. Cl_4S
B. $KClO_3$
C. Calcium dihydrogen phosphate
D. Sulfurous acid

A A is the answer because the atoms in the sulfur tetrachloride molecule are placed in order of increasing electronegativity. This formla is properly written as SCl_4. B is a proper formula for potassium chlorate. Calcium dihydrogen phosphate is $Ca(H_2PO_4)_2$. It derives its name from the Ca^{2+} cation in combination with an anion composed of a phosphate anion (PO_4^{3-}) that is doubly protonated to give a $H_2PO_4^-$ ion. Sulfurous acid is $H_2SO_3(aq)$.

70. Household "chlorine bleach" is sodium hypochlorite. Which of the following best represent the production of sodium hypochlorite, sodium chloride, and water by bubbling chlorine gas through aqueous sodium hydroxide?

A. $4Cl(g) + 4NaOH(aq) \rightarrow NaClO_2(aq) + 3NaCl(aq) + 2H_2O(l)$
B. $2Cl_2(g) + 4NaOH(aq) \rightarrow NaClO_2(aq) + 3NaCl(aq) + 2H_2O(l)$
C. $2Cl(g) + 2NaOH(aq) \rightarrow NaClO(aq) + NaCl(aq) + H_2O(l)$
D. $Cl_2(g) + 2NaOH(aq) \rightarrow NaClO(aq) + NaCl(aq) + H_2O(l)$

D. Chlorine gas is a diatomic molecule, eliminating choices A and C. The hypochlorite ion is ClO^- eliminating choices A and B. All of the equations are properly balanced.

71. Balance the equation for the neutralization reaction between phosphoric acid and calcium hydroxide by filling in the blank stoichiometric coefficients.

$$\_\_H_3PO_4 + \_\_Ca(OH)_2 \rightarrow \_\_Ca_3(PO_4)_2 + \_\_H_2O$$

A. 4, 3, 1, 4
B. 2, 3, 1, 8
C. 2, 3, 1, 6
D. 2, 1, 1, 2

C. We are given the unbalanced equation (**step 1**).

Next we determine the number of atoms on each side (**step 2**). For reactants (left of the arrow): 5H, 1P, 6O, and 1Ca. For products: 2H, 2P, 9O, and 3Ca.

We assume that the molecule with the most atoms—i.e. $Ca_3(PO_4)_2$—has a coefficient of one, and find the other coefficients required to have the same number of atoms on each side of the equation (**step 3**). Assuming $Ca_3(PO_4)_2$ has a coefficient of one means that there will be 3 Ca and 2 P on the right because H_2O has no Ca or P. A balanced equation would also have 3 Ca and 2 P on the left. This is achieved with a coefficient of 2 for H_3PO_4 and 3 for $Ca(OH)_2$. Now we have:

$$2H_3PO_4 + 3Ca(OH)_2 \rightarrow Ca_3(PO_4)_2 + ?H_2O$$

The coefficient for H_2O is found by a balance on H or on O. Whichever one is chosen, the other atom should be checked to confirm that a balance actually occurs. For H, there are 6 H from $2H_3PO_4$ and 6 from $3Ca(OH)_2$ for a total of 12 H on the left. There must be 12 H on the right for balance. None are accounted for by $Ca_3(PO_4)_2$, so all 12 H must occur on H_2O. It has a coefficient of 6.

$$2H_3PO_4 + 3Ca(OH)_2 \rightarrow Ca_3(PO_4)_2 + 6H_2O$$

This is choice C, but if time is available, it is best to check that the remaining atom is balanced. There are 8 O from $2H_3PO_4$ and 6 from $3Ca(OH)_2$ for a total of 14 on the left, and 8 O from $Ca_3(PO_4)_2$ and 6 from $6H_2O$ for a total of 14 on the right. The equation is balanced.

Mulitplication by a whole number (**step 4**) is not required because the stoichiometric coefficients from step 3 already are whole numbers.

An alternative method would be to try the coefficients given for answer A, answer B, etc. until we recognize a properly balanced equation.

72. Write an equation showing the reaction between calcium nitrate and lithium sulfate in aqueous solution. Include all products.

A. $CaNO_3(aq) + Li_2SO_4(aq) \rightarrow CaSO_4(s) + Li_2NO_3(aq)$
B. $Ca(NO_3)_2(aq) + Li_2SO_4(aq) \rightarrow CaSO_4(s) + 2LiNO_3(aq)$
C. $Ca(NO_3)_2(aq) + Li_2SO_4(aq) \rightarrow 2LiNO_3(s) + CaSO_4(aq)$
D. $Ca(NO_3)_2(aq) + Li_2SO_4(aq) + 2H_2O(l) \rightarrow 2LiNO_3(aq) + Ca(OH)_2(aq) + H_2SO_4(aq)$

B. When two ionic compounds are in solution, a precipitation reaction should be considered. We can determine from their names that the two reactants are the ionic compounds $Ca(NO_3)_2$ and Li_2SO_4. The compounds are present in aqueous solution as their four component ions Ca^{2+}, NO_3^-, Li^+, and SO_4^{2-}. Solubility rules indicate that nitrates are always soluble but sulfate will form a solid precipitate with Ca^{2+} forming $CaSO_4(s)$. Choice A results from assuming that the nitrate anion has a 2– charge instead of its 1– charge. B is correct. C assumes lithium nitrate is the precipitate. Choice D includes the reverse of a neutralization reaction. Water would not decompose due to the addition of these salts.

73. Find the mass of CO_2 produced by the combustion of 15 kg of isopropyl alcohol in the reaction:

$$2C_3H_7OH + 9O_2 \rightarrow 6CO_2 + 8H_2O$$

A. 33 kg
B. 44 kg
C. 50 kg
D. 60 kg

A Remember "grams to moles to moles to grams." Step 1 converts mass to moles for the known value. In this case, kg and kmol are used. Step 2 relates moles of the known value to moles of the unknown value by their stoichiometry coefficients. Step 3 converts moles off the unknown value to a mass.

$$15 \times 10^3 \text{ g } C_4H_8O \times \underbrace{\frac{1 \text{ mol } C_4H_8O}{60 \text{ g } C_4H_8O}}_{\text{step 1}} \times \underbrace{\frac{6 \text{ mol } CO_2}{2 \text{ mol } C_4H_8O}}_{\text{step 2}} \times \underbrace{\frac{44 \text{ g } CO_2}{1 \text{ mol } CO_2}}_{\text{step 3}} = 33 \times 10^3 \text{ g } CO_2$$

$$= 33 \text{ kg } CO_2$$

74. What is the density of nitrogen gas at STP? Assume an ideal gas and a value of 0.08206 L-atm/(mol-K) for the gas constant.

    A. 0.62 g/L
    B. 1.14 g/L
    C. 1.25 g/L
    D. 2.03 g/L

C The molecular mass M of N_2 is 28.0 g/mol.

$$d = \frac{nM}{V} = \frac{PM}{RT} = \frac{(1\,atm)\left(28.0\,\frac{g}{mol}\right)}{\left(0.08206\,\frac{L\cdot atm}{mol\cdot K}\right)(273.15\,K)} = 1.25\,\frac{g}{L}$$

Choice A results from forgetting that nitrogen is a diatomic gas. Choice B results from using a value of 25 °C for standard temperature. This is the thermodynamic standard temperature, but not STP.

A faster method is to recall that one mole of an ideal gas at STP occupies 22.4 L.

$$d\ (in\ \frac{g}{L}) = \frac{M\ (in\ \frac{g}{mol})}{22.4\,\frac{L}{mol}} = \frac{28.0\,\frac{g}{mol}}{22.4\,\frac{L}{mol}} = 1.25\,\frac{g}{L}.$$

75. Find the volume of methane that will produce 12 m³ of hydrogen in the reaction: $CH_4(g) + H_2O(g) \rightarrow CO(g) + 3H_2(g)$. Assume temperature and pressure remain constant.

    A. 4.0 m³
    B. 32 m³
    C. 36 m³
    D. 64 m³

A Stoichiometric coefficients may be used directly for ideal gas volumes at constant T and P because of Avogadro's Law.

$$12\ m^3\ H_2 \times \frac{1\ m^3\ CH_4}{3\ m^3\ H_2} = 4.0\ m^3\ CH_4$$

12 g of H_2 will be produced from 32 g of CH_4 (incorrect choice B).

76. A 100. L vessel of pure O_2 at 500. kPa and 20. °C is used for the combustion of butane:

$$2C_4H_{10} + 13O_2 \rightarrow 8CO_2 + 10H_2O.$$

Find the mass of butane to consume all the O_2 in the vessel. Assume O_2 is an ideal gas and use a value of R = 8.314 J/(mol•K).

A. 183 g
B. 467 g
C. 1.83 kg
D. 7.75 kg

A We are given a volume and asked for a mass. The steps will be "volume to moles to moles to mass."

"Volume to moles…" requires the ideal gas law, but first several units must be altered.

Units of joules are identical to m^3•Pa.
500 kPa is 500×10^3 Pa.
100 L is 0.100 m^3.
20 °C is 293.15 K.
$PV = nRT$ is rearranged to give:

$$n = \frac{PV}{RT} = \frac{\left(500 \times 10^3 \text{ Pa}\right)\left(0.100 \text{ m}^3 \text{ O}_2\right)}{\left(8.314 \dfrac{\text{m}^3 \cdot \text{Pa}}{\text{mol} \cdot \text{K}}\right)(293.15 \text{ K})} = 20.51 \text{ mol O}_2$$

"…to moles to mass" utilizes stoichiometry. The molecular weight of butane is 58.1 u.

$$20.51 \text{ mol O}_2 \times \frac{2 \text{ mol C}_4\text{H}_{10}}{13 \text{ mol O}_2} \times \frac{58.1 \text{ g C}_4\text{H}_{10}}{1 \text{ mol C}_4\text{H}_{10}} = 183 \text{ g C}_4\text{H}_{10}$$

77. Consider the reaction between iron and hydrogen chloride gas:

$$Fe(s) + 2HCl(g) \rightarrow FeCl_2(s) + H_2(g) .$$

7 moles of iron and 10 moles of HCl react until the limiting reagent is consumed. Which statements are true?

I.   HCl is the excess reagent
II. HCl is the limiting reagent
III. 7 moles of H_2 are produced
IV. 2 moles of the excess reagent remain

A. I and III
B. I and IV
C. II and III
D. II and IV

D The limiting reagent is found by dividing the number of moles of each reactant by its stoichiometric coefficient. The lowest result is the limiting reagent.

$$7 \text{ mol Fe} \times \frac{1 \text{ mol reaction}}{1 \text{ mol Fe}} = 7 \text{ mol reaction if Fe is limiting}$$

$$10 \text{ mol HCl} \times \frac{1 \text{ mol reaction}}{2 \text{ mol HCl}} = 5 \text{ mol reaction if HCl is limiting.}$$

Therefore, HCl is the limiting reagent (II is true) and Fe is the excess reagent.

5 moles of the reaction take place, so 5 moles of H_2 are produced, and of the 7 moles of Fe supplied, 5 are consumed, leaving 2 moles of the excess reagent (IV is true).

78. **32.0 g of hydrogen and 32.0 grams of oxygen react to form water until the limiting reagent is consumed. What is present in the vessel after the reaction is complete?**

    A. 16.0 g O_2 and 48.0 g H_2O
    B. 24.0 g H_2 and 40.0 g H_2O
    C. 28.0 g H_2 and 36.0 g H_2O
    D. 28.0 g H_2 and 34.0 g H_2O

C First the equation must be constructed:

$$2H_2 + O_2 \rightarrow 2H_2O$$

A fast and intuitive solution would be to recognize that:
1) One mole of H_2 is about 2.0 g, so about 16 moles of H_2 are present.
2) One mole of O_2 is 32.0 g, so one mole of is O_2 is present
3) Imagine the 16 moles of H_2 reacting with one mole of O_2. 2 moles of H_2 will be consumed before the one mole of O_2 is gone. O_2 is limiting. (Eliminate choice A.)
4) 16 moles less 2 leaves 14 moles of H_2 or about 28 g. (Eliminate choice B.)
5) The reaction began with 64.0 g total. Conservation of mass for chemical reactions forces the total final mass to be 64.0 g also. (Eliminate choice D.)

A more standard solution is presented next. First, mass is converted to moles:

$$32.0 \text{ g } H_2 \times \frac{1 \text{ mol } H_2}{2.016 \text{ g } H_2} = 15.87 \text{ mol } H_2 \quad \text{and} \quad 32.0 \text{ g } O_2 \times \frac{1 \text{ mol } O_2}{32.00 \text{ g } O_2} = 1.000 \text{ mol } O_2$$

Dividing by stoichiometric coefficients give

$$15.87 \text{ mol } H_2 \times \frac{1 \text{ mol reaction}}{2 \text{ mol } H_2} = 7.935 \text{ mol reaction if } H_2 \text{ is limiting}$$

$$1.000 \text{ mol } O_2 \times \frac{1 \text{ mol reaction}}{1 \text{ mol } O_2} = 1.000 \text{ mol reaction if } O_2 \text{ is limiting.}$$

O_2 is the limiting reagent, so no O_2 will remain in the vessel.

$$1.000 \text{ mol } O_2 \text{ consumed} \times \frac{2 \text{ mol } H_2O \text{ produced}}{1 \text{ mol } O_2} \times \frac{18.016 \text{ g } H_2O}{1 \text{ mol } H_2O} = 36.0 \text{ g } H_2O \text{ produced}$$

Rem

$$1.000 \text{ mol } O_2 \text{ consumed} \times \frac{2 \text{ mol } H_2 \text{ consumed}}{1 \text{ mol } O_2} \times \frac{2.016 \text{ g } H_2}{1 \text{ mol } H_2} = 4.03 \text{ g } H_2 \text{ consumed}$$

aining H_2 is found from:
32.0 g H_2 initially – 4.03 g H_2 consumed =28.0 g H_2 remain.

79. Three experiments were performed at the same initial temperature and pressure to determine the rate of the reaction

$$2ClO_2(g) + F_2(g) \rightarrow 2ClO_2F(g) .$$

Results are shown in the table below. Concentrations are given in millimoles per liter (mM).

Exp.	Initial $[ClO_2]$ (mM)	Initial $[F_2]$ (mM)	Initial rate of $[ClO_2F]$ increase (mM/sec)
1	5.0	5.0	0.63
2	5.0	20	2.5
3	10	10	2.5

What is the rate law for this reaction?

A. Rate $= k\left[F_2\right]$

B. Rate $= k\left[ClO_2\right]\left[F_2\right]$

C. Rate $= k\left[ClO_2\right]^2\left[F_2\right]$ — increase by 8

D. Rate $= k\left[ClO_2\right]\left[F_2\right]^2$

B   A four-fold increase in $[F_2]$ at constant $[ClO_2]$ between experiment one and two caused a four-fold increase in rate. Rate is therefore proportional to $[F_2]$ at constant $[ClO_2]$, eliminating choice D (Choice D predicts rate to increase by a factor of 16).

Between experiment 1 and 3, $[F_2]$ and $[ClO_2]$ both double in value. Once again, there is a four-fold increase in rate. If rate were only dependent on $[F_2]$ (choice A), there would be a two-fold increase. The correct answer, B, attributes a two-fold increase in rate to the doubling of $[F_2]$ and a two-fold increase to the doubling of $[ClO_2]$, resulting in a net four-fold increase. Choice C predicts a rate increase by a factor of 8.

If this were an elementary reaction describing a collision event between three molecules, choice C would be expected, but stoichiometry cannot be used to predict a rate law.

80. The reaction

$$(CH_3)_3CBr(aq) + OH^-(aq) \rightarrow (CH_3)_3COH(aq) + Br^-(aq)$$

occurs in three elementary steps:

$$(CH_3)_3CBr \rightarrow (CH_3)_3C^+ + Br^- \text{ is slow}$$
$$(CH_3)_3C^+ + H_2O \rightarrow (CH_3)_3COH_2^+ \text{ is fast}$$
$$(CH_3)_3COH_2^+ + OH^- \rightarrow (CH_3)_3COH + H_2O \text{ is fast}$$

What is the rate law for this reaction?

A. $\text{Rate} = k\left[(CH_3)_3CBr\right]$

B. $\text{Rate} = k\left[OH^-\right]$

C. $\text{Rate} = k\left[(CH_3)_3CBr\right]\left[OH^-\right]$

D. $\text{Rate} = k\left[(CH_3)_3CBr\right]^2$

A The first step will be rate-limiting. It will determine the rate for the entire reaction because it is slower than the other steps. This step is a unimolecular process with the rate given by answer A. Choice C would be correct if the reaction as a whole were one elementary step instead of three, but the stoichiometry of a reaction composed of multiple elementary steps cannot be used to predict a rate law.

81. Which statement about equilibrium is _not_ true?

A. Equilibrium shifts to minimize the impact of changes.
B. Forward and reverse reactions have equal rates at equilibrium.
C. A closed container of air and water is at a vapor-liquid equilibrium if the humidity is constant.
D. The equilibrium between solid and dissolved forms is maintained when salt is added to an unsaturated solution. — would be true for saturated solution

D Choice A is a restatement of Le Chatelier's Principle. B is a definition of equilibrium. A constant humidity (Choice C) occurs if the rate of vaporization and condensation are equal, indicating equilibrium. No solid is present in an **un**saturated solution. If solid is added, all of it dissolves indicating a lack of equilibrium. D would be true for a saturated soution.

82. Which statements about reaction rates are true?

 I. A catalyst will shift an equilibrium to favor product formation.
 II. Catalysts increase the rate of forward and reverse reactions.
 III. A greater temperature increases the chance that a molecular
      collision will overcome a reaction's activation energy.
 IV. A catalytic converter contains a homogeneous catalyst.

   A. I and II
   B. II and III
   C. II, III and IV
   D. I, III, and IV

B Catalysts provide an alternate mechanism in both directions, but do not alter
equilibrium (I is false, II is true). The kinetic energy of molecules increases
with temperature, so the energy of their collisions increases also (III is true).
Catalytic converters contain a _heterogeneous_ catalyst (IV is false).

83. Write the equilibrium expression K_{eq} for the reaction
    $CO_2(g) + H_2(g) \rightleftharpoons CO(g) + H_2O(l)$

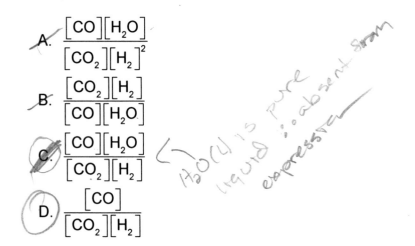

A. $\dfrac{[CO][H_2O]}{[CO_2][H_2]^2}$

B. $\dfrac{[CO_2][H_2]}{[CO][H_2O]}$

C. $\dfrac{[CO][H_2O]}{[CO_2][H_2]}$

D. $\dfrac{[CO]}{[CO_2][H_2]}$

D Product concentrations are multiplied together in the numerator and reactant
concentrations in the denominator, eliminating choice B. The stoichiometric
coefficient of H_2 is one, eliminating choice A. For heterogeneous reactions,
concentrations of pure liquids or solids are absent from the expression
because they are constant, eliminating choice C. D is correct.

84. What could cause this change in the energy diagram of a reaction?

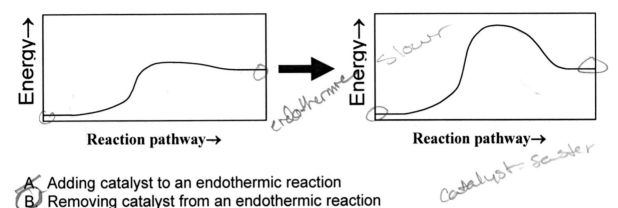

A. Adding catalyst to an endothermic reaction
B. Removing catalyst from an endothermic reaction
C. Adding catalyst to an exothermic reaction
D. Removing catalyst from an exothermic reaction

B The products at the end of the reaction pathway are at a greater energy than the reactants, so the reaction is endothermic (narrowing down the answer to A or B). The maximum height on the diagram corresponds to activation energy. An increase in activation energy could be caused by removing a heterogeneous catalyst.

85. $BaSO_4$ (K_{sp} = 1X10^{-10}) is added to pure H_2O. How much is dissolved in 1 L of saturated solution?

A. 2 mg
B. 10 μg
C. 2 μg
D. 100 pg

A $BaSO_4(s) \rightleftharpoons Ba^{2+}(aq) + SO_4^{2-}(aq)$, therefore: $K_{sp} = \left[Ba^{2+}\right]\left[SO_4^{2-}\right]$.

In a saturated solution: $\left[Ba^{2+}\right] = \left[SO_4^{2-}\right] = \sqrt{1 \times 10^{-10}} = 1 \times 10^{-5}$ M.

The mass in one liter is found from the molarity:

$$1 \times 10^{-5} \frac{\text{mol } Ba^{2+} \text{ or } SO_4^{2-}}{L} \times \frac{1 \text{ mol dissolved } BaSO_4}{1 \text{ mol } Ba^{2+} \text{ or } SO_4^{2-}} \times \frac{(137+32+4 \times 16)g \ BaSO_4}{1 \text{ mol } BaSO_4}$$

$$= 0.002 \frac{g}{L} \ BaSO_4 \times 1 \text{ L solution} \times \frac{1000 \text{ mg}}{g} = 2 \text{ mg } BaSO_4$$

86. The exothermic reaction $2NO(g) + Br_2(g) \rightleftharpoons 2NOBr(g)$ is at equilibrium. According to LeChatelier's principle:

   A. Adding Br_2 will increase [NO].
   B. An increase in container volume (with T constant) will increase [NOBr].
   C. An increase in pressure (with T constant) will increase [NOBr].
   D. An increase in temperature (with P constant) will increase [NOBr].

C LeChatelier's principle predicts that equilibrium will shift to partially offset any change. Adding Br_2 will be partially offset by reducing $[Br_2]$ and [NO] via a shift to the right (not choice A). For the remaining possibilities, we may write the reaction as: 3 moles $\rightleftharpoons$ 2 moles + heat. An increase in container volume will decrease pressure. This change will be partially offset by increasing the number of moles present, shifting the reaction to the left (not choice B). An increase in pressure will be offset by a decrease the number of moles present, shifting the reaction to the right (choice C, correct). Raising the temperature by adding heat will shift the reaction to the left (not choice D).

87. At a certain temperature, T, the equilibrium constant for the reaction $2NO(g) \rightleftharpoons N_2(g) + O_2(g)$ is $K_{eq} = 2 \times 10^3$. If a 1.0 L container at this temperature contains 90 mM N_2, 20 mM O_2, and 5 mM NO, what will occur?

   A. The reaction will make more N_2 and O_2.
   B. The reaction is at equilibrium.
   C. The reaction will make more NO.
   D. The temperature, T, is required to solve this problem.

A Calculate the reaction quotient at the actual conditions:

$$Q = \frac{[N_2][O_2]}{[NO]^2} = \frac{(0.090 \text{ M})(0.020 \text{ M})}{(0.005 \text{ M})^2} = 72$$

This value is less than K_{eq}: $72 < 2 \times 10^3$, therefore $Q < K_{eq}$. To achieve equilibrium, the numerator of Q must be larger relative to the denominator. This occurs when products turn into reactants. Therefore NO will react to make more N_2 and O_2.

88. Which statement about acids and bases is <u>not</u> true?

    A. All strong acids ionize in water.

    B. All Lewis acids accept an electron pair.

    C. All Brønsted bases use OH^- as a proton acceptor.

    D. All Arrhenius acids form H^+ ions in water.

C   Choice A is the definition of a strong acid, choice B is the definition of a Lewis acid, and choice D is the definition of an Arrhenius acid. By definition, all Arrhenius bases form OH^- ions in water, and all Brønsted bases are proton acceptors. But not all Brønsted bases use OH^- as a proton acceptor. NH_3 is a Brønsted base for example.

89. Which of the following are listed from weakest to strongest acid?

    A. H_2SO_3, H_2SeO_3, H_2TeO_3

    B. $HBrO$, $HBrO_2$, $HBrO_3$, $HBrO_4$

    C. HI, HBr, HCl, HF

    D. H_3PO_4, $H_2PO_4^-$, HPO_4^{2-}

B   The electronegativity of the central atom decreases from S to Se to Te as period number increases in the same periodic table group. The acidity of the oxide also decreases. Choice B is correct because acid strength increases with the oxidation state of the central atom. C is wrong because HI, HBr, and HCl are all strong acids but HF is a weak acid. D is wrong because acid strength is greater for polyprotic acids.

90. NH₄F is dissolved in water. Which of the following are conjugate acid/base pairs present in the solution?

~~I. NH₄⁺/NH₄OH~~
✗ II. HF/F⁻ ✓
III. H₃O⁺/H₂O ⎱ *always present in H₂O*
IV. H₂O/OH⁻ ⎰

A. I, II, and III
B. I, III, and IV
C. ~~II and IV~~
D. II, III, and IV ⟵ (circled)

D NH_4F is soluble in water and completely dissociates to NH_4^+ and F^-. F^- is a weak base with HF as its conjugate acid (II). NH_4^+ is a weak acid with NH_3 as its conjugate base. A conjugate acid/base pair must have the form HX/X (where X is one lower charge than HX). NH_4^+/NH_4OH (I) is not a conjugate acid/base pair, eliminating choice A and B. H_3O^+/H_2O and H_2O/OH^- (III and IV) are always present in water and in all aqueous solutions as conjugate acid/base pairs. All of the following equilibrium reactions occur in $NH_4F(aq)$:

$$NH_4^+(aq)+OH^-(aq)\rightleftharpoons NH_3(aq)+H_2O(l)$$
$$F^-(aq)+H_3O^+(aq)\rightleftharpoons HF(aq)+H_2O(l)$$
$$2H_2O(l)\rightleftharpoons H_3O^+(aq)+OH^-(aq)$$

91. What are the pH and the pOH of 0.010 M $HNO_3(aq)$?

A. pH = 1.0, pOH = 9.0
B. pH = 2.0, pOH = 12.0
C. pH = 2.0, pOH = 8.0
D. pH = 8.0, pOH = 6.0

B HNO_3 is a strong acid, so it completely dissociates:

$$\left[H^+\right]=0.010\ M=1.0\times10^{-2}\ M.$$
$$pH=-\log_{10}\left[H^+\right]=-\log_{10}\left(1.0\times10^{-2}\right)=2.0\ \text{(choices B or C)}.$$
$$\text{From }pH+pOH=14:\ pOH=12.0\ \text{(choice B)}.$$

92. What is the pH of a buffer made of 0.128 M sodium formate (HCOONa) and 0.072 M formic acid (HCOOH)? The pK_a of formic acid is 3.75.

   A. 2.0
   B. 3.0
   C. 4.0
   D. 5.0

C From the pK_a, we may find the K_a of formic acid:
$$K_a = 10^{-pK_a} = 10^{-3.75} = 1.78 \times 10^{-4}$$

This is the equilibrium constant:

$$K_a = \frac{[H^+][HCOO^-]}{[HCOOH]} = 1.78 \times 10^{-4} \text{ for the dissociation:}$$

$$HCOOH \rightleftharpoons H^+ + HCOO^-.$$

The pH is found by solving for the H^+ concentration:

$$[H^+] = K_a \frac{[HCOOH]}{[HCOO^-]} = (1.78 \times 10^{-4})\frac{0.072}{0.128} = 1.0 \times 10^{-4} \text{ M}$$

$$pH = -\log_{10}[H^+] = -\log_{10}(1.0 \times 10^{-4}) = 4.0 \text{ (choice C)}$$

93. A sample of 50.0 ml KOH is titrated with 0.100 M HClO$_4$. The initial buret reading is 1.6 ml and the reading at the endpoint is 22.4 ml. What is [KOH]?

   A. 0.0416 M
   B. 0.0481 M
   C. 0.0832 M
   D. 0.0962 M

A HClO$_4$ and KOH are both strong electrolytes. If you are good at memorizing formulas, solve the problem this way:

$$C_{unknown} = \frac{C_{known}(V_{final} - V_{initial})}{V_{unknown}} = \frac{0.100 \text{ M } (22.4 \text{ ml} - 1.6 \text{ ml})}{50.0 \text{ ml}} = 0.0416 \text{ M}.$$

The problem may also be solved by finding the moles of known substance:

$$0.100 \frac{mol}{L} \times \frac{1 L}{1000 \text{ mL}} \times (22.4 \text{ mL} - 1.6 \text{ mL}) = 0.00208 \text{ mol HClO}_4$$

This will neutralize 0.00208 mol KOH, and $\frac{0.00208 \text{ mol}}{0.0500 \text{ L}} = 0.0416 \text{ M}$

94. Rank the following from lowest to highest pH. Assume a small volume for the added component:

I.   0.01 mol HCl added to 1 L H_2O
II. 0.01 mol HI added to 1 L of an acetic acid/sodium acetate solution at pH 4.0
III. 0.01 mol NH_3 added to 1 L H_2O
IV. 0.1 mol HNO_3 added to 1 L of a 0.1 M $Ca(OH)_2$ solution

A. I < II < III < IV
B. I < II < IV < III
C. II < I < III < IV
D. II < I < IV < III

A   HCl is a strong acid. Therefore solution I has a <u>pH of 2</u> because

$$pH = -\log_{10}\left[H^+\right] = -\log_{10}\left(0.01\right) = 2.$$

HI is also a strong acid and would have a pH of 2 at this concentration in water, but the buffer will prevent pH from dropping this low. Solution II will have a pH <u>above 2</u> and below 4, eliminating choices C and D.

If a strong base were in solution III, its pOH would be 2. Using the equation pH + pOH = 14, its pH would be 12. Because NH_3 is a weak base, the pH of solution III will be greater than 7 and <u>less than 12</u>.

A neutralization reaction occurs in solution IV between 0.1 mol of H^+ from the strong acid HNO_3 and <u>0.2 mol of OH^-</u> from the strong base $Ca(OH)_2$. Each mole of $Ca(OH)_2$ contributes two base equivalents for the neutralization reaction. The base is the excess reagent, and 0.1 mol of OH^- remain after the reaction. This resulting solution will have a pOH of 1 and a <u>pH of 13</u>.

A is correct because: 2 < between 2 and 4 < between 7 and 12 < 13

95. The curve below resulted from the titration of a _____ _____ with a _____ _____ titrant.

A. weak acid, strong base
B. weak base, strong acid
C. strong acid, strong base
D. strong base, strong acid

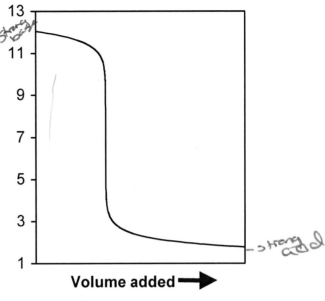

D The pH is above 7 initially and decreases, so an acid titrant is neutralizing a base. This eliminates A and C. The maximum slope (equivalence point) at the neutral pH of 7 indicates a strong base titrated with a strong acid, D.

96. Which statement about thermochemistry is true?

A. Particles in a system move about less freely at high entropy
B. Water at 100 °C has the same internal energy as water vapor at 100°C
C. A decrease in the order of a system corresponds to an increase in entropy.
D. At its sublimation temperature, dry ice has a higher entropy than gaseous CO_2

C At high entropy, particles have a large freedom of molecular motion (A is false). Water and water vapor at 100 °C contain the same translational kinetic energy, but water vapor has additional internal energy in the form of resisting the intermolecular attractions between molecules (B is false). We also know water vapor has a higher internal energy because heat must be added to boil water. Entropy may be thought of as the disorder in a system (C is correct). Sublimation is the phase change from solid to gas, and there is less freedom of motion for particles in solids than in gases. Solid CO_2 (dry ice) has a lower entropy than gaseous CO_2 because entropy decreases during a phase change that prevents molecular motion (D is false).

97. What is the heat change of 36.0 g H_2O at atmospheric pressure when its temperature is reduced from 125 °C to 40. °C? Use the following data:

A. −92.0 kJ
B. −10.8 kJ
C. 10.8 kJ
D. 92.0 kJ

Values for water	
Heat capacity of solid	37.6 J/mol•°C
Heat capacity of liquid	75.3 J/mol•°C
Heat capacity of gas	33.1 J/mol•°C
Heat of fusion	6.02 kJ/mol
Heat of vaporization	40.67 kJ/mol

A Heat is evolved from the substance as it cools, so the heat change will be negative, eliminating choices C and D. Data in the table are given using moles, so the first step is to convert the mass of water to moles:

$$36.0 \text{ g } H_2O \times \frac{1 \text{ mol } H_2O}{18.02 \text{ g } H_2O} = 2.00 \text{ mol } H_2O$$

There are three contributions to the heat evolved. First, the heat evolved when cooling the vapor from 125 °C to 100 °C is found from the heat capacity of the gas:

$$q_1 = n \times C \times \Delta T = 2.00 \text{ mol } H_2O(g) \times 33.1 \frac{J}{mol \, °C} \times (100 \text{ °C} - 125 \text{ °C})$$

$$= -1655 \text{ J to cool vapor}$$

Next, the heat evolved during condensation is found from the heat of vaporization:

$$q_2 = n \times (-\Delta H_{vaporization}) = 2.00 \text{ mol } H_2O \times (-40.67 \frac{kJ}{mol})$$

$$= -81.34 \text{ kJ to condense vapor}$$

Incorrect answer B results from using a heat of vaporization of 40.67 J/mol instead of kJ/mol.

Finally, the heat evolved when cooling the liquid from 100 °C to 40 °C is found from the heat capacity of the liquid:

$$q_3 = n \times C \times \Delta T = 2.00 \text{ mol } H_2O(g) \times 75.3 \frac{J}{mol \, °C} \times (40 \text{ °C} - 100 \text{ °C})$$

$$= -9036 \text{ J to cool liquid}$$

The total heat change is the sum of these contributions:

$$q = q_1 + q_2 + q_3 = -1.655 \text{ kJ} + (-81.34 \text{ kJ}) + (-9.036 \text{ kJ}) = -92.03 \text{ kJ}$$

$$= -92.0 \text{ kJ (Choice A)}$$

98. **What is the standard heat of combustion of $CH_4(g)$? Use the following data:**

A. −890.3 kJ/mol
B. −604.5 kJ/mol
C. −252.9 kJ/mol
D. −182.5 kJ/mol

Standard heats of formation	
$CH_4(g)$	−74.8 kJ/mol
$CO_2(g)$	−393.5 kJ/mol
$H_2O(l)$	−285.8 kJ/mol

A First we must write a balanced equation for the combustion of CH_4. The balanced equation is:

$$CH_4(g) + 2O_2(g) \rightarrow CO_2(g) + 2H_2O(l).$$

The heat of combustion may be found from the sum of the productions minus the sum of the reactants of the heats of formation:

$$\Delta H_{rxn} = H_{product\ 1} + H_{product\ 2} + \ldots - \left(H_{reactant\ 1} + H_{reactant\ 2} + \ldots\right)$$

$$= \Delta H_f^{\circ}(CO_2) + 2\Delta H_f^{\circ}(H_2O) - \left(\Delta H_f^{\circ}(CH_4) + 2\Delta H_f^{\circ}(O_2)\right)$$

The heat of formation of an element in its most stable form is zero by definition, so $\Delta H_f^{\circ}(O_2(g)) = 0 \ \dfrac{kJ}{mol}$, and the remaining values are found from the table:

$$\Delta H_{rxn} = -393.5 \ \frac{kJ}{mol} + 2(-285.8 \ \frac{kJ}{mol}) - \left(-74.8 \ \frac{kJ}{mol} + 2(0)\right) = -890.3 \ \frac{kJ}{mol} \ \text{(choice A)}$$

99. **Which reaction creates products at a lower total entropy than the reactants?**

A. Dissolution of table salt: $NaCl(s) \rightarrow Na^+(aq) + Cl^-(aq)$
B. Oxidation of iron: $4Fe(s) + 3O_2(g) \rightarrow 2Fe_2O_3(s)$
C. Dissociation of ozone: $O_3(g) \rightarrow O_2(g) + O(g)$
D. Vaporization of butane: $C_4H_{10}(l) \rightarrow C_4H_{10}(g)$

B Choice A is incorrect because two particles are at a greater entropy than one and because ions in solution have more freedom of motion than a solid. For B (the correct answer), the products are at a lower entropy than the reactants because there are fewer product molecules and they are all in the solid form but one of the reactants is a gas. Reaction B is still spontaneous because it is highly exothermic. For C, there are more product molecules than reactants, and for D, the gas phase is always at a higher entropy than the liquid.

100. Which statement about reactions is true?

A. All spontaneous reactions are both exothermic and cause an increase in entropy.
B. An endothermic reaction that increases the order of the system cannot be spontaneous.
C. A reaction can be non-spontaneous in one direction and also non-spontaneous in the opposite direction.
D. Melting snow is an exothermic process

B All reactions that are both exothermic and cause an increase in entropy will be spontaneous, but the converse (choice A) is not true. Some spontaneous reactions are exothermic but decrease entropy and some are endothermic and increase entropy. Choice B is correct. The reverse reaction of a non-spontaneous reaction (choice C) will be spontaneous. Melting snow (choice D) requires heat. Therefore it is an endothermic process

101. 10. kJ of heat are added to one kilogram of Iron at 10. °C. What is its final temperature? The specific heat of iron is 0.45 J/g•°C.

A. 22 °C
B. 27 °C
C. 32 °C
D. 37 °C

C The expression for heat as a function of temperature change:

$$q = n \times C \times \Delta T$$

may be rearranged to solve for the temperature change:

$$\Delta T = \frac{q}{n \times C}.$$

In this case, n is a mass and C is the specific heat of iron:

$$\Delta T = \frac{10000 \text{ J}}{1000 \text{ g} \times 0.45 \ \frac{\text{J}}{\text{g} \ °\text{C}}} = 22 \ °\text{C}.$$

This is not the final temperature (choice A is incorrect). It is the temperature difference between the initial and final temperature.

$$\Delta T = T_{final} - T_{initial} = 22 \ °\text{C}$$

Solving for the final temperature gives us:

$$T_{final} = \Delta T + T_{initial} = 22 \ °\text{C} + 10 \ °\text{C} = 32 \ °\text{C} \text{ (Choice C)}$$

102. Which reaction is <u>not</u> a redox process?

 A. Combustion of octane: $2C_8H_{18} + 25O_2 \rightarrow 16CO_2 + 18H_2O$

 B. Depletion of a lithium battery: $Li + MnO_2 \rightarrow LiMnO_2$

 C. Corrosion of aluminum by acid: $2Al + 6HCl \rightarrow 2AlCl_3 + 3H_2$

 D. Taking an antacid for heartburn:
  $CaCO_3 + 2HCl \rightarrow CaCl_2 + H_2CO_3 \rightarrow CaCl_2 + CO_2 + H_2O$

D The oxidation state of atoms is altered in a redox process. During combustion (choice A), the carbon atoms are oxidized from an oxidation number of –4 to +4. Oxygen atoms are reduced from an oxidation number of 0 to –2. All batteries (choice B) generate electricity by forcing electrons from a redox process through a circuit. Li is oxidized from 0 in the metal to +1 in the $LiMnO_2$ salt. Mn is reduced from +4 in manganese(IV) oxide to +3 in lithium manganese(III) oxide salt. Corrosion (choice C) is due to oxidation. Al is oxidized from 0 to +3. H is reduced from +1 to 0. Acid-base neutralization (choice D) transfers a proton (an H atom with an oxidation state of +1) from an acid to a base. The oxidation state of all atoms remains unchanged (Ca at +2, C at +4, O at –2, H at +1, and Cl at –1), so D is correct. Note that choices C and D both involve an acid. The availability of electrons in aluminum metal favors electron transfer but the availability of CO_3^{2-} as a proton acceptor favors proton transfer.

103. Given the following heats of reaction:

$$\Delta H = -0.3 \text{ kJ / mol for} \quad Fe(s) + CO_2(g) \rightarrow FeO(s) + CO(g)$$

$$\Delta H = 5.7 \text{ kJ / mol for} \quad 2Fe(s) + 3CO_2(g) \rightarrow Fe_2O_3(s) + 3CO(g)$$

and $\Delta H = 4.5 \text{ kJ / mol for} \quad 3FeO(s) + CO_2(g) \rightarrow Fe_3O_4(s) + CO(g)$

use Hess's Law to determine the heat of reaction for:

$$3Fe_2O_3(s) + CO(g) \rightarrow 2Fe_3O_4(s) + CO_2(g)?$$

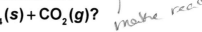

make reactant

A. −10.8 kJ/mol
B. −9.9 kJ/mol
C. −9.0 kJ/mol
D. −8.1 kJ/mol

B We are interested in $3Fe_2O_3$ as a reactant. Only the second reaction contains this molecule, so we will take three times the opposite of the second reaction. We are interested in $2Fe_3O_4$ as a product, so we will take two times the third reaction. An intermediate result is:

$3Fe_2O_3(s) + 9CO(g) \rightarrow 6Fe(s) + 9CO_2(g)$ $\Delta H = -3 \times 5.7 \text{ kJ/mol} = -17.1 \text{ kJ/mol}$

$6FeO(s) + 2CO_2(g) \rightarrow 2Fe_3O_4(s) + 2CO(g)$ $\Delta H = 2 \times 4.5 \text{ kJ/mol} = 9.0 \text{ kJ/mol}$

$3Fe_2O_3(s) + 6FeO(s) + 7CO(g) \rightarrow$

$\quad 2Fe_3O_4(s) + 6Fe(s) + 7CO_2(g)$ $\Delta H = (-17.1 + 9.0) \text{ kJ/mol} = -8.1 \text{ kJ/mol}$

However, D is not the correct answer because it is not ΔH for the reaction of the problem statement. We may use six times the first reaction to eliminate both FeO and Fe from the intermediate result and obtain the reaction of interest:

$3Fe_2O_3(s) + 6FeO(s) + 7CO(g) \rightarrow$

$\quad 2Fe_3O_4(s) + 6Fe(s) + 7CO_2(g)$ $\Delta H = -8.1 \text{ kJ/mol}$

$6Fe(s) + 6CO_2(g) \rightarrow 6FeO(s) + 6CO(g)$ $\Delta H = 6 \times (-0.3 \text{ kJ/mol}) = -1.8 \text{ kJ/mol}$

$3Fe_2O_3(s) + CO(g) \rightarrow 2Fe_3O_4(s) + CO(g)$

$\Delta H = (-8.1 + -1.8) \text{ kJ/mol}$
$\quad = -9.9 \text{ kJ/mol (choice B)}$

104. What is the oxidant in the reaction: $2H_2S + SO_2 \rightarrow 3S + 2H_2O$?

   A. H_2S
   B. SO_2
   C. S
   D. H_2O

B   The S atom in H_2S has an oxidation number of -2 and is oxidized by SO_2 (the oxidant, choice B) to elemental sulfer (oxidation number $= 0$). The S atom in SO_2 has an oxidation number of $+4$ and is reduced. The two half-reactions are:

$$SO_2 + 4e^- + 4H^+ \xrightarrow{\text{reduction}} S + 2H_2O$$

$$2H_2S \xrightarrow{\text{oxidation}} 2S + 4e^- + 4H^+$$

105. Molten NaCl is subjected to electrolysis. What reaction takes place at the cathode?

   A. $2Cl^-(l) \rightarrow Cl_2(g) + 2e^-$
   B. $Cl_2(g) + 2e^- \rightarrow 2Cl^-(l)$
   C. $Na^+(l) + e^- \rightarrow Na(l)$
   D. $Na^+(l) \rightarrow Na(l) + e^-$

C   Reduction (choices B and C) always occurs at the cathode. Molten NaCl is composed of ions in liquid form before electrolysis (answer C). A and D are oxidation reactions, and D is also not properly balanced because a $+1$ charge is on the left and a -1 charge is on the right. The two half-reactions are:

$$Na^+(l) + e^- \xrightarrow{\text{reduction at cathode}} Na(l)$$

$$2Cl^-(l) \xrightarrow{\text{oxidation at anode}} Cl_2(g) + 2e^-$$

The net reaction is:
$$2NaCl(l) \rightarrow 2Na(l) + Cl_2(g)$$

106. What is the purpose of the salt bridge in a voltaic cell?

    A. To receive electrons from the oxidation half-reaction
    B. To relieve the buildup of positive charge in the anode half-cell
    C. To conduct electron flow
    D. To permit positive ions to flow from the cathode half-cell to the anode half-cell

D The anode receives electrons from the oxidation half-reaction (choice A) and the circuit conducts electron flow (choice C) to the cathode which supplies electrons for the reduction half-reaction. This flow of electrons from the anode to the cathode is relieved by a flow of ions through the salt bridge from the cathode to the anode (answer D). The salt bridge relieves the buildup of positive charge in the cathode half-cell (choice B is incorrect).

107. Given $E°=-2.37$ V for $Mg^{2+}(aq)+2e^-\rightarrow Mg(s)$ and $E°=0.80$ V for $Ag^+(aq)+e^- \rightarrow Ag(s)$, what is the standard potential of a voltaic cell composed of a piece of magnesium dipped in a 1 M Ag^+ solution and a piece of silver dipped in 1 M Mg^{2+}?

    A. 0.77 V
    B. 1.57 V
    C. 3.17 V
    D. 3.97 V

C $Ag^+(aq)+e^-\rightarrow Ag(s)$ has a larger value for $E°$ (reduction potential) than $Mg^{2+}(aq)+2e^-\rightarrow Mg(s)$. Therefore, in the cell described, reduction will occur at the Ag electrode and it will be the cathode. Using the equation:

$$E^o_{cell} = E^o(cathode) - E^o(anode), \text{ we obtain:}$$

$$E^o_{cell} = 0.80 \text{ V} - (-2.37 \text{ V}) = 3.17 \text{ V (Answer C)}.$$

Choice D results from the incorrect assumption that electrode potentials depend on the amount of material present. The balanced net reaction for the cell is:

$$Mg(s) \rightarrow Mg^{2+}(aq) + 2e^- \qquad\qquad E^o_{ox} = 2.37 \text{ V}$$

$$\underline{2Ag^+(aq) + 2e^- \rightarrow 2Ag(s)} \qquad\qquad \underline{E^o_{red} = 0.80 \text{ V (not 1.60 V)}}$$

$$Mg(s) + 2Ag^+(aq) \rightarrow 2Ag(s) + Mg^{2+}(aq) \qquad E^o_{cell} = 3.17 \text{ V (not 3.97 V)}$$

108. A proper name for this hydrocarbon is:

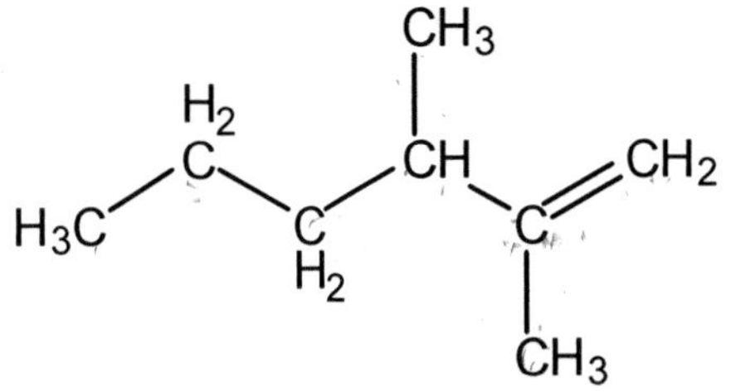

A. 4,5-dimethyl-6-hexene
B. 2,3-dimethyl-1-hexene
C. 4,5-dimethyl-6-hexyne
D. 2-methyl-3-propyl-1-butene

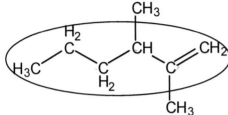

B The hydrocarbon contains a double bond and no triple bonds, so it is an alkene. Choice C describes an alkyne. The longest carbon chain is six carbons long, corresponding to a parent molecule of 1-hexene (circled to the left). Choice D is an improper name because it names the molecule as a substituted butane, using a shorter chain as the parent molecule.

Finally, the lowest possible set of locant numbers must be used. Choice A is an improper name because the larger possible set of locant numbers is chosen.

109. An IUPAC approved name for this molecule is:

A. butanal
B. propanal
C. butanoic acid
D. propanoic acid

3 aldehydes
3 carboxylic

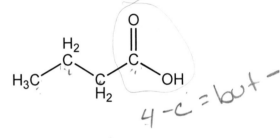

4-c = but-

C The COOH group means that the molecule is a carboxylic acid and its name will use the suffix –oic acid. The presence of 4 carbon atoms means the prefix butan- will be used. An alternate name for the molecule is butyric acid. Choices A and B would be used for aldehydes (CHO group). Choices B and D would be used for 3 carbon atoms:

butanal (also called butyraldehyde):

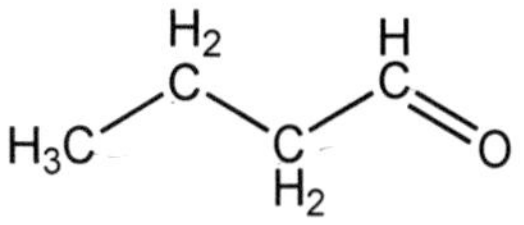

propanal (also called propionaldehyde):

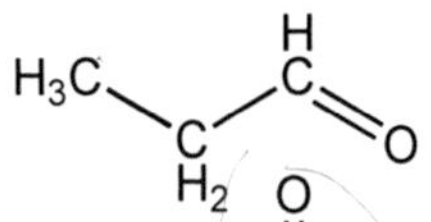

propanoic acid (also called propionic acid):

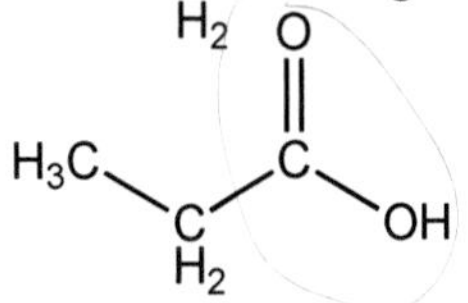

110. Which molecule has a systematic name of methyl ethanoate? *esters*

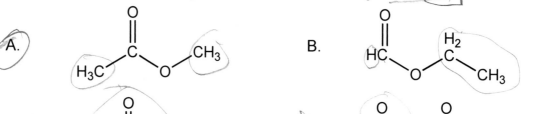

A.

B.

C. 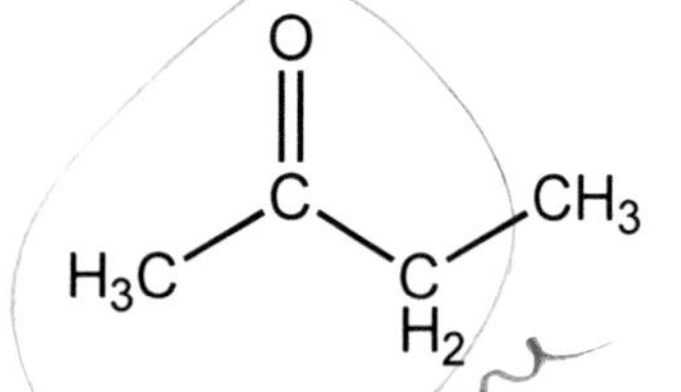 *ketone*

D.

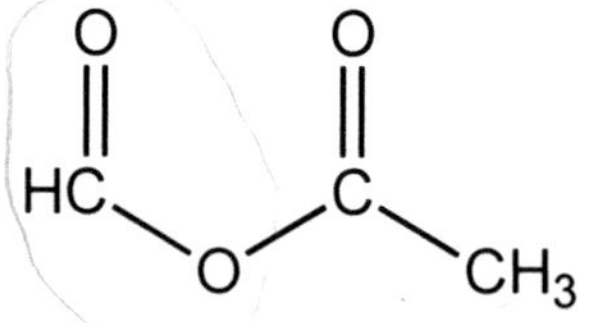

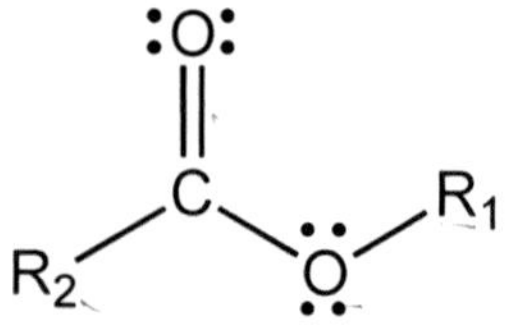

A   The suffix *–oate* is used for esters. The ester group is shown to the right. Choice C is a ketone (ethyl methyl ketone or 2-butanone). The ketone group is shown to the left. Choice D is an acid anhydride (ethanoic methanoic anhydride). The acid anhydride group is shown below to the right. A and B are both esters. The hydrocarbon R_2 with the carbonyl group receives the *–oate* suffix and the hydrocarbon R_1 with the *-yl* suffix is attached to the other oxygen. Choice B is ethyl methanoate and A is correct.

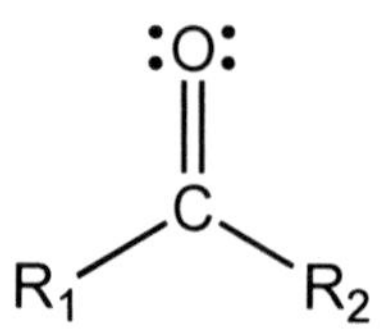

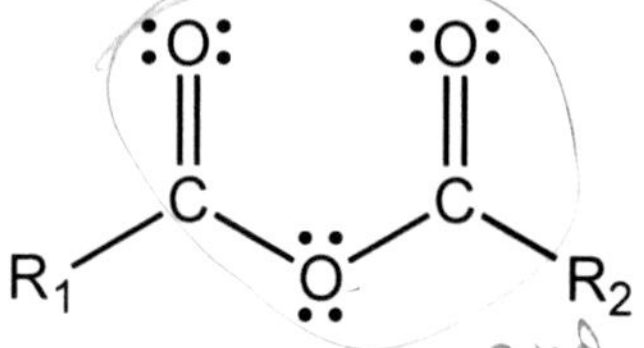

anhydrd

111. This compound contains an:

A. alkene, carboxylic acid, ester, and ketone
B. aldehyde, alkyne, ester, and ketone
C. aldehyde, alkene, carboxylic acid, and ester
D. acid anhydride, aldehyde, alkene, and amine

C The derivatives are circled below:

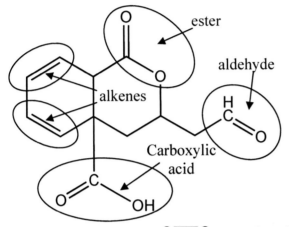

Choice A is wrong because there are no ketones in the molecule. A ketone has a carbonyl group linked to two hydrocarbons as shown to the right. All the carbonyls in the molecule are linked to at least one oxygen atom. Choice B is wrong because there are no ketones and no alkynes in the molecule. An alkyne contains a C≡C triple bond. Choice D is wrong because there are no acid anhydrides (shown to the left) and no amines (shown to the right). Amines require at least one N-C bond and there are no nitrogen atoms in the molecule.

112. Which group of scientists made contributions in the same area of chemistry?

A. Volta, Kekulé, Faraday, London
B. Hess, Joule, Kelvin, Gibbs
C. Boyle, Charles, Arrhenius, Pauli
D. Davy, Mendeleev, Ramsay, Galvani

B Hess, Joule, Kelvin, and Gibbs all contributed to thermochemistry and have thermodynamic entities named after them. Volta, Faraday, and Galvani (choice D) contributed to electrochemistry, Kekulé to organic chemistry, London to chemical bonding, Boyle and Charles to gas laws, Arrhenius to acid/base chemistry and thermochemstry, Pauli to quantum theory, Davy and Ramsay to element isolation, and Mendeleev to the periodic table.

113. Which of the following pairs are isomers?

I.

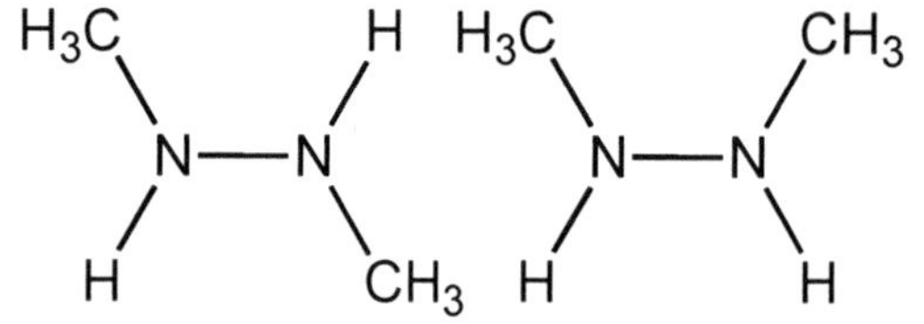

II. pentanal 2-pentanone

III.

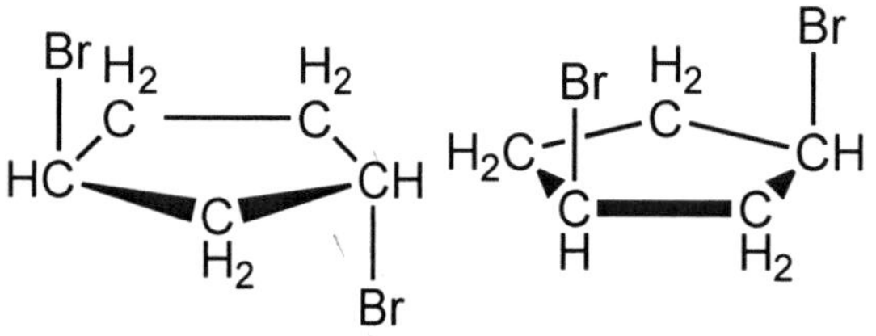

IV.

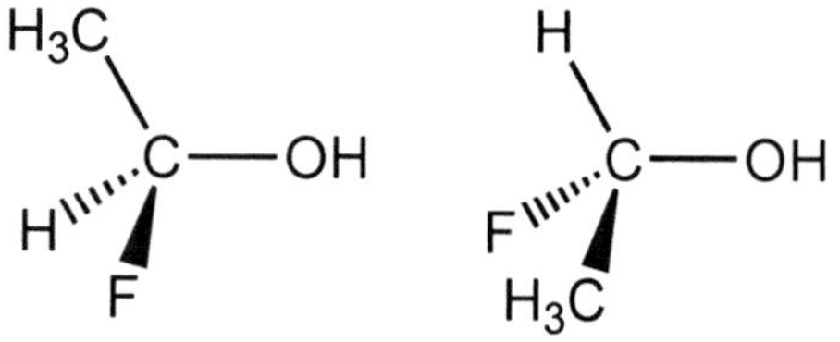

A. I and IV
B. II and III
C. I, II, and III
D. I, II, III, and IV

B In pair I, the N—N bond may freely rotate in the molecule because it is not a double bond. The identical molecule is represented twice.

For pair II, pentanal is

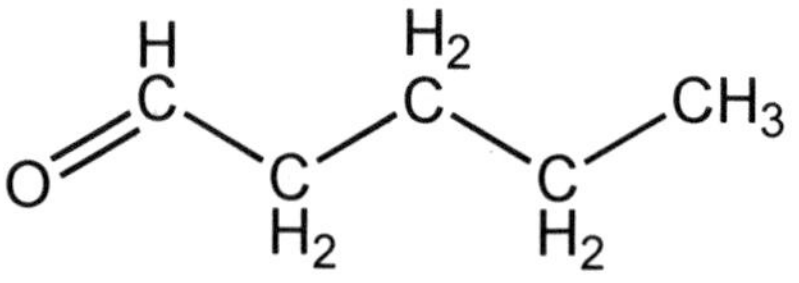

and 2-pentanone is:

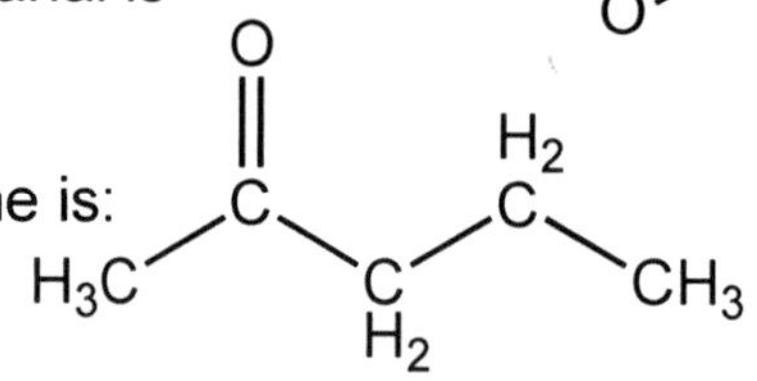

Both molecules are $C_5H_{10}O$, and they are isomers because they have the same formula with a different arrangement of atoms.

In pair III, both molecules are 1,3-dibromocyclopentane, $C_5H_8Br_2$. In the first molecule, the bromines are in a *trans* configuration, and in the second molecule, they are *cis*. The two molecules are also viewed from different perspectives. Unlike pair I, no bond rotation may occur because the intervening atoms are locked into place by the ring, so they are different arrangements and are isomers.

In pair IV (1-fluoroethanol), there is a chiral center, so stereoisomers are possible, but as in pair I, the same molecule is represented twice. Rotating the C-O bond indicates that the two structures are superimposable. This molecule: to the right is a stereoisomer to the molecule represented in IV. The answer is B (pairs II and III).

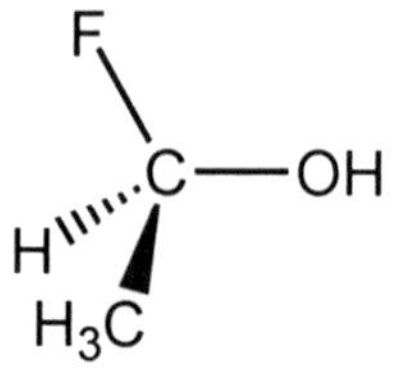

114. Which instrument would be most useful for separating two different proteins from a mixture?

A. UV/Vis spectrophotometer
B. Mass spectrometer
C. Gas chromatograph
D. Liquid chromatograph

D UV/Vis spectrophotometry measures the light at ultraviolet and visible wavelengths that can pass through the mixture, and mass spectrometry determines molecular weights. Both might be used to find the concentration of each protein, but neither is a separation technique. Gas chromatography is used for small molecules in the gas phase. Proteins are too large to exist in the gas phase. Liquid chromatography (answer D) is used to separate large molecules.

115. Classify these biochemicals.

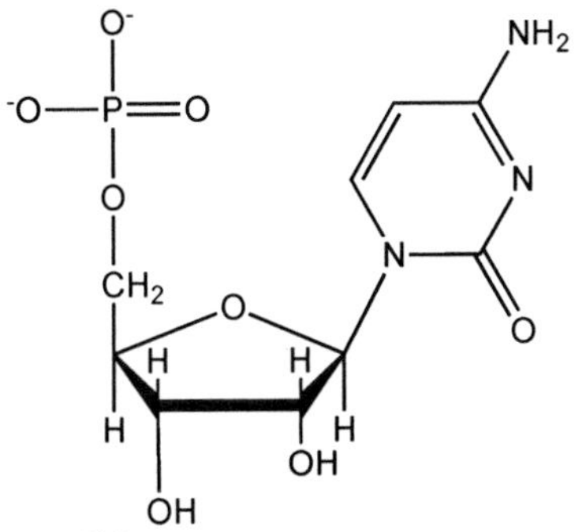

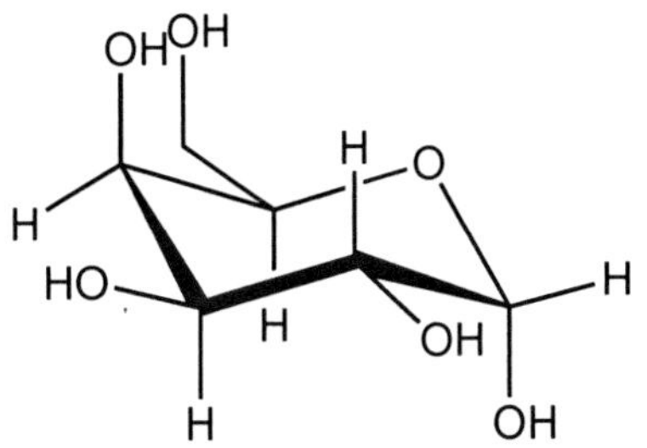

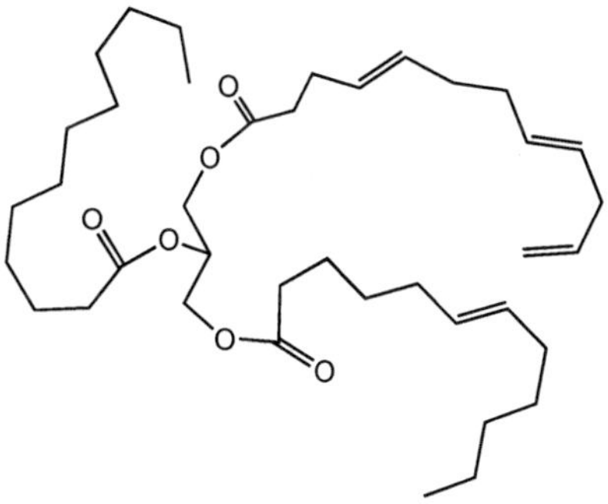

A. I-nucleotide, II-sugar, III-peptide, IV-fat
B. I-disaccharide, II-sugar, III-fatty acid, IV-polypeptide
C. I-disaccharide, II-amino acid, III-fatty acid, IV- polysaccharide
D. I-nucleotide, II-sugar, III-triacylglyceride, and IV-DNA

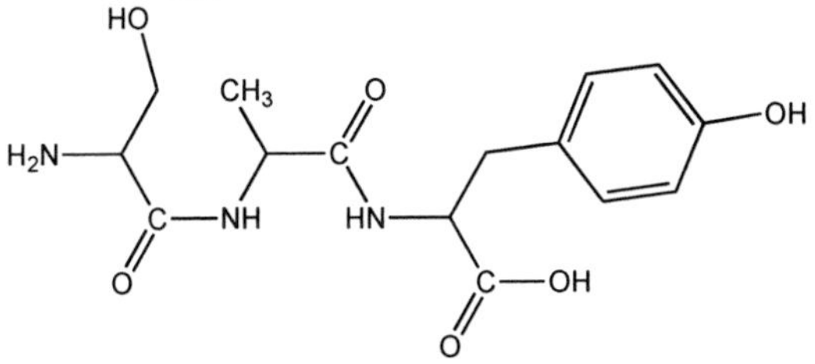

A I is a phosphate (PO_4) linked to a sugar and an amine: a nucleotide. II has the formula $C_nH_{2n}O_n$, indicative of a sugar. III contains three amino acids linked with peptide bonds. It is a tripeptide. IV is a triacylglyceride, a fat molecule.

116. You create a solution of 2.00 µg/ml of a pigment and divide the solution into 12 samples. You give four samples each to three teams of students. They use a spectrophotometer to determine the pigment concentration. Here is their data:

Team	Concentration (µg/ml)			
	sample 1	sample 2	sample 3	sample 4
1	1.98	1.93	1.92	1.88
2	1.70	1.72	1.69	1.70
3	1.78	1.99	2.87	2.20

← more precise

Which of the following are true?

A. Team 1 has the most precise data
B. Team 3 has the most accurate data in spite of it having low precision
C. The data from team 2 is characteristic of a systematic error
D. The data from team 1 is more characteristic of random error than the data from team 3.

C For choice A, the data from team 2 are closer to the mean for team 2 than the data from team 1 are to its mean. Therefore, team 1's data does not have the most precision.

For choice B, the mean from team 1 is near 1.9 µg/ml (we don't need to calculate exact values). It differs from the actual value by 0.1 µg/ml. The mean from team 2 is near 1.7 µg/ml and is inaccurate by 0.3 µg/ml. The mean from team 3 is not obvious, but it may be calculated as 2.21 µg/ml, differing from the actual value by about 0.2 µg/ml. Team 3's data is less accurate than the data from team 1.

The data from team 2 is clustered close to a central value but this value is wrong. Low accuracy with high precision is indicative of a systematic error. (C is correct).

For choice D, a lack of precision is indicative of random error, and the data from team 1 is more precise than the data from team 3.

117. Which pair of measurements have an identical meaning?

A. 32 micrometers and 0.032 g
B. 26 nm and 2.60×10^{-8} m
C. 3.01×10^{-5} m^3 and 30.1 ml
D. 0.0020 L and 20 cm^3

C For A, the prefix *micro*— indicates 10^{-6}. 32 micrograms is 0.000032 g. For B, the two measurements do not have the same meaning because they differ in the number of significant figures. 26 nm is 2.6×10^{-8} m. The symbol "n" for *nano*— indicates 10^{-9}. For C and D, unit conversions between cubic meters and liters are required.

$$\text{For C: } 3.01 \times 10^{-5} \text{ m}^3 \times \frac{1000 \text{ L}}{1 \text{ m}^3} \times \frac{1000 \text{ ml}}{1 \text{ L}} = 30.1 \text{ ml (C is correct)}.$$

$$\text{For D: } 0.0020 \text{ L} \times \frac{1 \text{ m}^3}{1000 \text{ L}} \times \frac{(100)^3 \text{ cm}^3}{1 \text{ m}^3} = 2.0 \text{ cm}^3 \text{ (D is incorrect)}.$$

118. Match the instrument with the quantity it measures

I. eudiometer
II. calorimeter
III. manometer
IV. hygrometer

A. I-volume, II-mass, III-radioactivity, IV-humidity
B. I-volume, II-heat, III-pressure, IV-humidity
C. I-viscosity, II-mass, III-pressure, IV-surface tension
D. I-viscosity, II-heat, III-radioactivity, IV-surface tension

B A eudiometer is a straight tube used to measure gas volume by liquid exclusion. A calorimeter is a device used to measure changes in heat. A manometer is a U-shaped tube used to measure pressure. A hygrometer measures humidity (Answer B). Mass is measured with a balance, radioactivity is measured with a Geiger counter or scintillation counter. Viscosity is measured with a viscometer, surface tension is measured by several different techniques.

119. Four nearly identical gems from the same mineral are weighed using different balances. Their masses are:

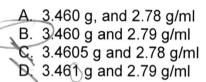

3.4533 g, 3.459 g, 3.4656 g, 3.464 g.

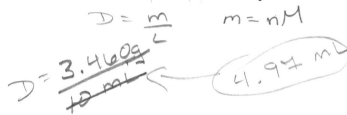

The four gems are then collected and added to a volumetric cylinder containing 10.00 ml of liquid, and a new volume of 14.97 ml is read. What is the average mass of the four stones and what is the density of the mineral?

A. 3.460 g, and 2.78 g/ml
B. 3.460 g and 2.79 g/ml
C. 3.4605 g and 2.78 g/ml
D. 3.461 g and 2.79 g/ml

B The average mass is the sum of the four readings divided by four:

(3.4533 g + 3.459 g + 3.4656 g + 3.464 g)/4 = 3.460475 g (caculator value)

This value must be rounded off to three significant digits <u>after the decimal point</u> because this is the lowest precision of the added values. The four is an exact number. This means rounding downwards to 3.460 g, eliminating choices C and D. The volume of the collected stones is found from the increase in the level read off the cylinder:

$$14.97 \text{ ml} - 10.00 \text{ ml} = 4.97 \text{ ml}$$

The density is found by dividing the sum of the masses by this volume:

$$\frac{3.4533 \text{ g} + 3.459 \text{ g} + 3.4656 \text{ g} + 3.464 \text{ g}}{4.97 \text{ ml}} = \frac{13.8419 \text{ g}}{4.97 \text{ ml}} = 2.7850905 \text{ g/ml (caculator value)}$$

This value must be rounded off to three <u>total</u> significant digits because this is the lower precision of the numerator and the denominator. The first insignificant digit is a 5. In this case there are additional non-zero digits after the 5, so rounding occurs upwards to 2.79 g/ml (answer B).

120. Which list includes equipment that would not be used in vacuum filtration.

    A. Rubber tubing, Florence flask, Büchner funnel
    B. Vacuum pump, Hirsch funnel, rubber stopper with a single hole
    C. Aspirator, filter paper, filter flask
    D. Lab stand, clamp, filter trap

A Florence flasks are round-bottomed and are used for uniform heating. They do not have the hose barb or the thick wall needed to serve as a filter flask during vacuum filtration. Only a designated filter flask should be used during vacuum filtration. Every other piece of equipment could be used in filtration. A spatula is often used to scrape dried product off of filter paper.

121. Which of the following statements about lab safety is not true?

    A. Corrosive chemicals should be stored below eye level.
    B. A chemical splash on the eye or skin should be rinsed for 15 minutes in cold water.
    C. MSDS means "Material Safety Data Sheet."
    D. A student should "stop, drop, and roll" if their clothing catches fire in the lab.

D In the lab, the safety shower should be used.

122. Which of the following lists consists entirely of chemicals that are considered safe enough to be in a high school lab?

    A. hydrochloric acid, lauric acid, potassium permanganate, calcium hydroxide
    B. ethyl ether, nitric acid, sodium benzoate, methanol
    C. cobalt (II) sulfide, ethylene glycol, benzoyl peroxide, ammonium chloride
    D. picric acid, hydrofluoric acid, cadmium chloride, carbon disulfide.

A Hydrochloric acid (HCl) is a common acid reagent in high school chemistry. Lauric acid is the fatty acid $CH_3(CH_2)_{10}COOH$ also known as dodecanoic acid. Potassium permanganate ($KMnO_4$) is a strong oxidizer. Calcium hydroxide ($Ca(OH)_2$) is a strong base. These chemicals in their pure state are hazardous, but they are considered safe enough to be in high schools. Ethyl ether (Choice B) should not be in high schools because it may form highly explosive organic peroxides over time. Benzoyl peroxide (choice C) at low concentrations in gel form is an acne medication, but the pure compound is highly explosive. Choice D consists entirely of chemicals that are too dangerous for high schools. Picric acid is highly explosive, hydrofluoric acid is very corrosive and very toxic, all cadmium compounds are highly toxic, and carbon disulfide is explosive and toxic.

123. The following procedure was developed to find the specific heat capacity of metals:

1. Place pieces of the metals in an ice-water bath so their initial temperature is 0 °C.
2. Weigh a styrofoam cup.
3. Add water at room temperature to the cup and weigh it again
4. Add a cold metal from the bath to the cup and weigh the cup a third time.
5. Monitor the temperature drop of the water until a final temperature at thermal equilibrium is found.

_____ is also required as additional information in order to obtain heat capacities for the metals. The best control would be to follow the same protocol except to use _____ in step 4 instead of a cold metal.

A. The heat capacity of water / a metal at 100 °C
B. The heat of formation of water / ice from the 0 °C bath
C. The heat of capacity of ice / glass at 0 °C
D. The heat capacity of water / water from the 0 °C bath

D The equation:

$$q = n \times C \times \Delta T$$

is used to determine what additional information is needed. The specific heat, C, of the metals may be found from the heat added, the amount of material, and the temperature change. The amount of metal is found from the difference in weight between steps 3 and 4, and the temperature change is found from the difference between the final temperature and 0 °C. The additional value required is the heat added, q. This may be found from the heat removed from the water if the amount of water, the heat capacity of water, and the temperature change of water are known. The amount of water is found from the difference in weight between step 2 and 3, and the temperature change is found from the difference between the final temperature and room temperature. The only additional information required is the heat capacity of water, eliminating choices B and C. Heat of formation (choice B) is only used for chemical reactions.

A good control simplifies only the one aspect under study without adding anything new. Metal at 100 °C (choice A) would alter the temperature of the experiment and glass (choice C) would add an additional material to the study. Ice (choice B) would require consideration of the heat of fusion. Choice D is an ideal control because the impact of water at 0 °C on room temperature water is simpler than the impact of metals at 0 °C on room temperature water, and nothing new is added.

124. Which statement about the impact of chemistry on society is **not** true?

   A. Partial hydrogenation creates *trans* fat.
   B. The Haber Process incorporates nitrogen from the air into molecules for agricultural use.
   C. The CO_2 concentration in the atmosphere has decreased in the last ten years.
   D. The concentration of ozone-destroying chemicals in the stratosphere has decreased in the last ten years.

C CO_2 concentrations in the atmosphere continue to increase (answer C), but the concentration of ozone destroying chemicals has fallen (answer D) due to international agreements.

125. Which statement about everyday applications of chemistry is **true**?

   A. Rainwater found near sources of air pollution will most likely be basic.
   B. Batteries run down more quickly at low temperatures because chemical reactions are proceeding more slowly.
   C. Benzyl alcohol is a detergent used in shampoo.
   D. Adding salt decreases the time required for water to boil.

B Sources of air pollution (choice A) will most likely cause acid rain.

Low temperatures decrease reaction rates, and this is also true of electrochemical reactions in batteries. At low temperature, less current is supplied and the effect will be a short life for applications that demand current. (Answer B is correct).

Benzyl alcohol (choice C) has the formula shown to the right. Like detergents, this molecule has a non-polar region (the benzene ring) and a polar region (the hydroxyl group). But, unlike detergents, the non-polar region for benzyl alcohol is small and short. Detergents have long, "tail-like" non-polar regions that can surround oils and grease. Benzyl alcohol is sometimes included in shampoo to prevent itching and bacterial growth.

Adding salt (choice D) increases the boiling point of water, thus increasing the time required for water to boil. It decreases the time required to cook food once boiling occurs.

Sample Open-Response Questions

Directions: Read the information below and complete the given exercise. Explain your reasoning and show your work.

126. Level: Challenging.

A teapot containing 675 g of water at 25.0 °C is placed on a kitchen stove and heated to 100.0 °C until just before it begins to boil. Natural gas is delivered to the stove at a rate of 135 mL per second at 25.0 °C and a constant total pressure of 1.13 atm. Natural gas is supplied with the following composition:

Weight percentage		
Methane	Ethane	Carbon dioxide
94.9%	4.4 %	0.7%

a. How much energy is required to heat the water in the teapot? The specific heat of water is 4.18 J/(g$\cdot$°C).

b. Write balanced equations for the combustion of methane and of ethane.

c. How many moles of gas are supplied to the stove each second? Assume that the gas behaves as an ideal gas.
$R = 0.08205$ L$\cdot$atm/(mol$\cdot$K).

d. How many moles of methane and ethane are supplied to the stove each second?

e. How much heat is produced by hydrocarbon combustion each second? Assume complete combustion and use the following values:

Heat of combustion (kJ/mol)	
Methane	Ethane
890	2900

f. What is the mass of carbon dioxide released into the atmosphere each second? Assume complete combustion.

g. Using only the answers from A and E, estimate a length of time for the water to reach 100.0 °C. An experiment was performed and the water was observed to reach 100.0 °C in 258 seconds. Provide a reason why this value differs from the estimated value.

127. Level: Intermediate.

Students are learning about equilibrium in a chemistry laboratory exercise. The relevant materials include: an aqueous solution of $CoCl_2$, concentrated HCl, water, a hot water bath, an ice-water bath, and all necessary glassware. $CoCl_2$ dissociates in water into Cl^- ions and Co^{2+} ions which form the hydrated complex $[Co(H_2O)_6]^{2+}$. This ionic complex turns the solution pink. When HCl is added, the additional Cl^- reacts with $[Co(H_2O)_6]^{2+}$ in a mildly endothermic reversible reaction to form $[CoCl_4]^{2-}$. This ion turns the solution blue.

Write an essay describing a qualitative (not quantitative) investigation to explore the effects of both concentration and temperature on equilibrium. In your essay:

a. Describe an appropriate experimental design.

b. Describe the kind of data that will need to be gathered and how the data will be analyzed.

c. Describe the expected results of the study and relate the results to the reactions involved and the relevant concepts of chemical equilibrium.

Sample Open-Response Answers

Note: Many chemistry essays on certification exams consist of quantitative problem solving with the requirement to show your work. The first of the two sample essays is of this type. For additional practice, I recommend solving quantitative problems from the multiple choice sample test with an "essay mindset" and comparing your essays to the solutions shown in the "Answers with Solutions" section. Some chemistry essays require little or no quantitative problem solving, but they ask for an experimental design or analysis of a design. The second of the two sample essays is of this type. These essays usually have no single correct solution.

126.

A) The energy required to heat the water in the teapot may be found from the mass, specific heat, and temperature change of the water by utilizing the expression:

$$q = m \times C \times \Delta T \quad \text{where } q \Rightarrow \text{heat added}$$

$$m \Rightarrow \text{mass of water}$$

$$C \Rightarrow \text{specific heat of water}$$

$$\Delta T \Rightarrow \text{change in temperature } T_{final} - T_{initial}$$

Substituting values yields:

$$q = 675 \text{ g} \times 4.18 \, \frac{J}{g \cdot °C} \times (100.0 \text{ °C} - 25.0 \text{ °C}) = 211 \times 10^3 \text{ J}$$

$$= 211 \text{ kJ}$$

211 kJ of energy are required.

B) The chemical formula for methane is CH_4 and the formula for ethane is C_2H_6. During combustion reactions, the substance reacts with oxygen, and products consist of compounds with oxygen with each atom at its highest possible oxidation state. For C, this product is CO_2, and for H it is H_2O. The unbalanced equations are:

$$CH_4 + O_2 \rightarrow CO_2 + H_2O \quad \text{for methane and}$$

$$C_2H_6 + O_2 \rightarrow CO_2 + H_2O \quad \text{for ethane}$$

The most complex molecule in both cases is the hydrocarbon, and so a stoichiometric coefficient of one will be assumed for now.

For methane, this results in 1 C atom on both the left and right, so C is balanced. There are 4 H atoms on the left side of the equation and 2 on the right. A stoichiometric coefficient of 2 for H_2O corrects this imbalance:

$$CH_4 + ?O_2 \rightarrow CO_2 + 2H_2O .$$

Finally, there are 2 O atoms on the left and 4 on the right. A stoichiometric coefficient of 2 for O_2 balances the equation:

$$CH_4 + 2O_2 \rightarrow CO_2 + 2H_2O .$$

For ethane, there are 2 C atoms on the left and one on the right, so a coefficient of 2 will initially be given to CO_2. There are 6 H atoms on the left and 2 on the right, so H_2O will have a coefficient of 3:

$$C_2H_6 + ?O_2 \rightarrow 2CO_2 + 3H_2O .$$

There are 2 O atoms on the left and 7 on the right. A fractional stoichiometric coefficient describes the combustion of one mole of ethane:

$$C_2H_6 + \frac{7}{2}O_2 \rightarrow 2CO_2 + 3H_2O$$

Finally, the fractional coefficient could be eliminated by multiplying the entire expression by 2:

$$2C_2H_6 + 7O_2 \rightarrow 4CO_2 + 6H_2O$$

A final check confirms that there are now 4 C atoms, 12 H atoms, and 14 O atoms on both sides of the equation.

C) The problem states that 135 mL of an ideal gas are supplied to the stove each second. The pressure and temperature are also known. The ideal gas equation, $PV = nRT$, may be rearranged to solve for the number of moles of gas flowing in a second:

$$n = \frac{PV}{RT} .$$

135 mL is converted to 0.135 L to correspond to the given units of the ideal gas constant. . 25.0 °C is converted to Kelvin before using the ideal gas law:

$$273.15 + 25.0 = 298.15 \text{ K (the last digit isn't significant)}$$

Plugging these values into the equation for one second of gas flow yields:

$$n = \frac{1.13 \text{ atm} \times 0.135 \ \dfrac{L}{s}}{0.08205 \ \dfrac{L \cdot atm}{mol \cdot K} \times 298.15 \text{ K}} = 6.24 \times 10^{-3} \ \frac{mol}{s} .$$

0.00624 moles of gas are supplied to the stove each second.

D) Weight percentages of methane and ethane must first be converted to mole fractions using the molecular weights of all three species. These fractions will then be used with the answer to part C to determine the number of moles of each hydrocarbon supplied to the stove every second.

The molecular weights of the three components are:

For methane: $12.011 + 4 \times 1.0079 = 16.043$ g/mole CH_4

For ethane: $2 \times 12.011 + 6 \times 1.0079 = 30.069$ g/mole C_2H_6 .

For carbon dioxide: $12.011 + 2 \times 15.999 = 44.009$ g/mole CO_2

The molecular weights and weight percentages given in the table will be used to find the number of moles of each component using a basis of exactly 1 g of natural gas:

$$\frac{0.949 \text{ g } CH_4}{\text{g gas}} \times \frac{\text{mole } CH_4}{16.043 \text{ g } CH_4} = \frac{0.05915 \text{ mole } CH_4}{\text{g gas}}$$

$$\frac{0.044 \text{ g } C_2H_6}{\text{g gas}} \times \frac{\text{mole } C_2H_6}{30.069 \text{ g } C_2H_6} = \frac{0.00146 \text{ mole } C_2H_6}{\text{g gas}} .$$

$$\frac{0.007 \text{ g } CO_2}{\text{g gas}} \times \frac{\text{mole } CO_2}{44.009 \text{ g } CO_2} = \frac{0.00016 \text{ mole } CO_2}{\text{g gas}}$$

These intermediate results contain an extra, insignificant digit. The three values above are added together to give the total number of moles in a gram of gas:

$$0.05915 + 0.00146 + 0.00016 = 0.06077 \frac{\text{mole gas}}{\text{g gas}} .$$

The mole fractions of the hydrocarbon components may then be found:

$$\frac{0.05915 \text{ mole } CH_4}{0.06077 \text{ mole gas}} = 0.9733 \frac{\text{mole } CH_4}{\text{mole gas}}$$

$$\frac{0.00146 \text{ mole } C_2H_6}{0.06077 \text{ mole gas}} = 0.0240 \frac{\text{mole } C_2H_6}{\text{mole gas}} .$$

Finally, these values are multiplied by the result from part C to give the moles of hydrocarbons supplied each second. The extra insignificant digit for the ethane problem is removed from this final result.

$$0.00624 \frac{\text{mole gas}}{\text{s}} \times 0.9733 \frac{\text{mole } CH_4}{\text{mole gas}} = 0.00607 \frac{\text{mole } CH_4}{\text{s}}$$

$$0.00624 \frac{\text{mole gas}}{\text{s}} \times 0.0240 \frac{\text{mole } C_2H_6}{\text{mole gas}} = 0.00015 \frac{\text{mole } C_2H_6}{\text{s}} .$$

E). The heats of combustion are multiplied by the rate of supply for each gas:

$$0.00607 \, \frac{\text{mole } CH_4}{s} \times 890 \, \frac{kJ}{\text{mole } CH_4} = 5.4 \, \frac{kJ}{s} \text{ from methane combustion}$$

$$0.00015 \, \frac{\text{mole } C_2H_6}{s} \times 2900 \, \frac{kJ}{\text{mole } C_2H_6} = 0.44 \, \frac{kJ}{s} \text{ from ethane combustion}$$

The total heat produced from hydrocarbon combustion is the sum of these two values: 5.8 kJ each second.

F) There are three sources of carbon dioxide in this problem. CO_2 already in the natural gas before combustion is released into the atmosphere. This value is found from its weight percentage and values calculated in parts C and D:

$$\frac{0.007 \text{ g } CO_2}{\text{g gas}} \times \frac{\text{g gas}}{0.06077 \text{ mole gas}} \times \frac{0.00624 \text{ mole gas}}{s} = 0.0007 \frac{\text{g } CO_2}{s}$$

CO_2 from combustion is found from the values calculated in part D and the stoichiometry of the chemical equations from part B.

For methane:
$$0.00607 \, \frac{\text{mole } CH_4}{s} \times \frac{1 \text{ mole } CO_2}{1 \text{ mole } CH_4} \times \frac{44.009 \text{ g } CO_2}{1 \text{ mole } CO_2} = 0.267 \frac{\text{g } CO_2}{s}.$$

For ethane:
$$0.00015 \, \frac{\text{mole } C_2H_6}{s} \times \frac{4 \text{ mole } CO_2}{2 \text{ mole } C_2H_6} \times \frac{44.009 \text{ g } CO_2}{1 \text{ mole } CO_2} = 0.013 \frac{\text{g } CO_2}{s}.$$

The mass of CO_2 released is found from the sum of these three contributions:

$$0.0007 + 0.267 + 0.013 = 0.281 \frac{\text{g } CO_2}{s}.$$

G) From A, 211 kJ are required to heat the water. From E, the rate of heat produced by combustion is 5.8 kJ per second. An estimate of the number of seconds to heat the water may be found by dividing the heat required by the rate at which it is supplied:

$$211 \, kJ \times \frac{1 \text{ s}}{5.8 \text{ kJ}} = 36 \text{ seconds}.$$

One reason why this value differs from the observed value of 258 seconds is because the heat supplied by combustion does not transfer perfectly into the heating of water in an insulated, adiabatic process. Heat from combustion will also be used to raise the temperature of the teapot, the stovetop, and nearby air. Heat from the hot water is also lost to the air.

127. The dissociation of $CoCl_2$ in water and the formation of the pink-colored complex is described by the reaction:

$$CoCl_2(s) + 6H_2O(l) \rightarrow [Co(H_2O)_6]^{2+}(aq)_{PINK} + 2Cl^-(aq)$$

The dissociation of HCl is described by the reaction:

$$HCl(aq) \rightarrow H^+(aq) + Cl^-(aq).$$

When these two solutions are combined, the pink solution is expected to turn blue due to the formation of $[CoCl_4]^{2-}$. The equilibrium reaction under study is:

$$[Co(H_2O)_6]^{2+}(aq)_{PINK} + 4Cl^-(aq) + heat \rightleftharpoons [CoCl_4]^{2-}(aq)_{BLUE} + 6H_2O(l)$$

The relevant concept under study is Le Chatelier's principle and its application to the impact of concentration and temperature on equilibrium.

The only data that will be gathered in this qualitative study is an observation of color changes. Every color change should be recorded in the students' lab notebooks. A color change from pink to blue indicates the reaction above is occurring from left to right. A color change from blue to pink indicates the reaction is occurring from right to left.

The first step will provide a large volume of uniform experimental material for the students. This step will utilize concentrated HCl, so it should be performed by the instructor while the students watch. The instructor should wear gloves. Concentrated HCl should be slowly added to the cobalt(II) chloride solution until the entire solution changes color. Students should note the color before and after the change takes place. This should be a change from pink to blue because the equilibrium reaction under study has shifted to the right to partially offset the impact of added chloride ion. A sufficient volume should be prepared to provide every student or team of students with an aliquot of 10 mL of this blue solution. Even though these volumes are small, the students are still handling a corrosive acid at moderate concentration and should wear gloves to minimize their risk of contact. The students should perform the following procedure and answer the following questions:

1. Label six test tubes 1 through 6.
2. Place half of your blue solution in tube 1 and half in tube 2.
3. Add water to tube 2 until a change in color takes place. Which direction does the equilibrium shift when water is added? (Answer: Adding water shifted equilibrium from right to left)
4. Divide the solution in tube 1 in half. Place half in tube 3 and half in tube 4.

5. Divide the solution in tube 2 in half. Place half in tube 5 and half in tube 6.
6. Test tube 3 and 4 should contain blue solution and the solution in tubes 5 and 6 should be pink.
7. Place tubes 3 and 5 in the hot water bath. Which solution changes color? (Answer: solution in tube 5 turns blue) Which direction did the equilibrium shift? (Answer: Adding heat shifted equilibrium from left to right).
8. Place tubes 4 and 6 in the ice-water bath. Which test tube changes color? (Answer: solution in tube 4 turns pink) Which direction did the equilibrium shift? (Answer: Removing heat shifted equilibrium from right to left).

The following table summarizes the experimental study in terms of predictions from Le Chatelier's principle and the expected results in the lab:

$$[Co(H_2O)_6]^{2+}(aq)_{PINK} + 4Cl^-(aq) + heat \rightleftharpoons [CoCl_4]^{2-}(aq)_{BLUE} + 6H_2O(l)$$

Test tube	Treatment	Prediction	Result
1	Instructor added Cl⁻	Shift to the right	Pink to blue (moved to 3 and 4)
2	Add H_2O	Shift to the left	Blue to pink (moved to 5 and 6)
3	Add heat	Shift to the right	No change (or deeper blue)
4	Remove heat	Shift to the left	Blue to pink
5	Add heat	Shift to the right	Pink to blue
6	Remove heat	Shift to the left	No change (or deeper pink)

XAMonline, INC. 21 Orient Ave. Melrose, MA 02176

Toll Free number 800-509-4128

TO ORDER Fax 781-662-9268 OR www.XAMonline.com

MASSACHUSETTS TEST FOR EDUCATOR LICENTURE - MTEL - 2007

PO#   Store/School:

Address 1:

Address 2 (Ship to other):

City, State Zip

Credit card number_____-_____-_____-_____   expiration_____

EMAIL _____

PHONE   **FAX**

13# ISBN 2007	TITLE	Qty	Retail	Total
978-1-58197-884-1	MTEL Biology 13			
978-1-58197-883-4	MTEL Chemistry 12			
978-1-58197-875-9	MTEL Communication and Literacy Skills 01			
978-1-58197-885-8	MTEL Earth Science 14			
978-1-58197-879-7	MTEL English 07			
978-1-58197-892-6	MTEL Foundations of Reading 90 (requirement all El. Ed)			
978-1-58197-887-2	MTEL French 26			
978-1-58197-876-6	MTEL General Curriculum (formerly Elementary) 03			
978-1-58197-877-3	MTEL General Curriculum (formerly Elementary) 03 Sample Questions			
978-1-58197-881-0	MTEL General Science 10			
978-1-58197-878-0	MTEL History 06 (Social Science)			
978-1-58197-196-5	MTEL Latin & Classical Humanities 15			
978-1-58197-880-3	MTEL Mathematics 09			
978-1-58197-890-2	MTEL Middle School Humanities 50			
978-1-58197-889-6	MTEL Middle School Mathematics 47			
978-1-58197-891-9	MTEL Middle School Mathematics-Science 51			
978-1-58197-886-5	MTEL Physical Education 22			
978-1-58197-882-7	MTEL Physics Sample Test 11			
978-1-58197-898-8	MTEL Political Science/Political Philosophy 48			
978-1-58197-888-9	MTEL Spanish 28			
			SUBTOTAL	
	FOR PRODUCT PRICES VISIT WWW.XAMONLINE.COM		**Ship**	$8.25
			TOTAL	